T0219763

Lecture Notes in Computer Science 9209

Commenced Publication in 1973
Founding and Former Series Editors:
Gerhard Goos, Juris Hartmanis, and Jan van Leeuwen

More information about this series at http://www.springer.com/series/7409

Balder ten Cate · Alessandra Mileo (Eds.)

Web Reasoning and Rule Systems

9th International Conference, RR 2015
Berlin, Germany, August 4–5, 2015
Proceedings

 Springer

Editors
Balder ten Cate
University of California
Santa Cruz
USA

Alessandra Mileo
National University of Ireland
Galway
Ireland

ISSN 0302-9743   ISSN 1611-3349   (electronic)
Lecture Notes in Computer Science
ISBN 978-3-319-22001-7   ISBN 978-3-319-22002-4   (eBook)
DOI 10.1007/978-3-319-22002-4

Library of Congress Control Number: 2015944501

LNCS Sublibrary: SL3 – Information Systems and Applications, incl. Internet/Web, and HCI

Printed on acid-free paper

Springer International Publishing AG Switzerland is part of Springer Science+Business Media
(www.springer.com)

Preface

Advances in Semantic Web and linked data research and standardization have established formats and technologies for representing, sharing, and re-using knowledge on the Web. The scale, velocity, and heterogeneous nature of Web data however, poses many challenges, and turns basic tasks such as query answering, data transformations, and knowledge integration into complex reasoning problems. Rule-based systems and rule-based extensions to Web languages have found numerous applications where the ability to synthesize relevant knowledge from noisy, distributed, heterogeneous, dynamic, incomplete, and possibly contradicting information is key.

The International Conference on Web Reasoning and Rule Systems has become a major forum for discussion and dissemination of new results on relevant topics in these areas, spanning research areas from computational intelligence and agent-based systems to Web technologies and information extraction.

This volume contains the proceedings of the 9th International Conference on Web Reasoning and Rule Systems (RR 2015), held during August 4–5, 2015, in Berlin, Germany. The conference program included presentations of five full research papers and four technical communications, a more concise paper format that provides the opportunity to present preliminary and ongoing work, systems, and applications that are of interest to the RR audience. The conference program also included keynote talks by Michael Genesereth, Benny Kimelfeld, Lora Aroyo, and Riccardo Rosati, covering both the scientific and the industrial perspectives of Web reasoning and rule systems. Extended abstracts of these talks are included in this volume. This year's edition of RR was organized in conjunction with the International Web Rule Symposium (RuleML), and the talk by Michael Genesereth was a joint keynote with RuleML. The accepted papers were selected out of 16 submissions, of which 11 were full papers and five were technical communications. Each submission received at least three reviews. Five full papers and two technical communications were accepted, and two full papers were accepted as technical communications.

In order to foster the engagement of students and their precious contribution to the research community, RR also hosted a doctoral consortium and a joint poster session. The participation and contribution of students to the RR poster session have also been fostered by the established co-location with the 11th edition of the Reasoning Web Summer School, held just before RR.

We want to thank the invited speakers for their valuable contribution, and the local organizer Adrian Paschke and his team who did a magnificent job with the organization of the event. We would like to thank our general chair Wolfgang Faber, our doctoral consortium chair Marco Montali, and our publicity chair Luca Pulina. We also thank Marco Maratea for his very successful work as a sponsorship chair, and we gratefully acknowledge the support of our sponsors. As usual, EasyChair provided a fantastic support as conference management system and for the preparation of these proceedings. Last but not least, we thank all authors and participants of RR 2015, the essential constituents of the scientific community; we hope they had an enjoyable stay in Berlin.

May 2015   Balder ten Cate
                                                               Alessandra Mileo

Organization

Program Committee

Darko Anicic	Siemens AG, Germany
Marcelo Arenas	PUC, Chile
Marcello Balduccini	Drexel University, USA
Leopoldo Bertossi	Carleton University, Canada
Meghyn Bienvenu	CNRS and Université Paris-Sud, France
Fernando Bobillo	University of Zaragoza, Spain
Daniel Deutch	Tel Aviv University, Israel
Agostino Dovier	Università di Udine, Italy
Thomas Eiter	Technical University of Vienna (TU Wien), Austria
Sergio Flesca	DEIS - University of Calabria, Italy
Paul Fodor	Stony Brook University, USA
Andre Freitas	University of Passau, Germany
André Hernich	University of Liverpool, UK
Stijn Heymans	Sheldonize, LLC, USA
Aidan Hogan	DCC, Universidad de Chile, Chile
Benny Kimelfeld	Technion, Israel and LogicBlox, USA
Roman Kontchakov	Birkbeck, University of London, UK
Markus Krötzsch	Technische Universität Dresden, Germany
Georg Lausen	University of Freiburg, Germany
Joohyung Lee	Arizona State University, USA
Domenico Lembo	Sapienza University of Rome, Italy
Carsten Lutz	Universität Bremen, Germany
Thomas Meyer	University of Cape Town and CAIR, South Africa
Alessandra Mileo	INSIGHT Centre for Data Analytics, National University of Ireland, Galway, Ireland
Marco Montali	Free University of Bozen-Bolzano, Italy
Boris Motik	University of Oxford, UK
Marie-Laure Mugnier	University of Montpellier (LIRMM/Inria), France
Matthias Nickles	National University of Ireland, Galway and INSIGHT Centre for Data Analytics, Ireland
Giorgio Orsi	University of Oxford, UK
Magdalena Ortiz	Vienna University of Technology, Austria
Jeff Z. Pan	University of Aberdeen, UK
Adrian Paschke	Freie Universität Berlin, Germany
Axel Polleres	Vienna University of Economics and Business – WU Wien, Austria
Lucian Popa	IBM Research - Almaden, USA

Francesco Ricca   Department of Mathematics and Computer Science,
                                University of Calabria, Italy
Riccardo Rosati   DIAG, Sapienza Università di Roma, Italy
Sebastian Rudolph   Technische Universität Dresden, Germany
Luciano Serafini   Fondazione Bruno Kessler, Italy
Evgeny Sherkhonov   University of Amsterdam, The Netherlands
Steffen Staab   University of Koblenz-Landau, Germany and
                                University of Southampton, UK
Umberto Straccia   ISTI-CNR, Italy
Balder ten Cate   UC Santa Cruz, USA

Additional Reviewers

Dao-Tran, Minh
Gao, Feng
Havur, Giray
Leblanc, Emily
Milani, Mostafa
Nguyen, Hai H.
Zhao, Yuting

The generous sponsors of RR 2015 include:

    Association for Logic Programming

    Artificial Intelligence

   European Coordinating Committee for
                        Artificial Intelligence

    LogicBlox

   Siemens

Invited Talks

Truth Is a Lie: Rules and Semantics from Crowd Perspectives

Lora Aroyo

VU University Amsterdam and Tagasauris, Inc.
lora.aroyo@vu.nl

Processing real-world data with the crowd leaves one thing absolutely clear — there is no single notion of truth, but rather a spectrum that has to account for context, opinions, perspectives and shades of grey. CrowdTruth is a new framework for processing of human semantics drawn more from the notion of consensus than from set theory.

The quest for truth is the main drive behind research in the whole spectrum of humanities and sciences. It is well accepted that statements about music, art, politics, etc. have a range of interpretations and perspectives, and cannot easily be treated with a simple binary notion of truth. Is Obama a good president? Did Picasso paint his masterpieces in his blue period? It is not as well accepted that scientific statements in domains such as medicine, physics, etc. have the same feature. Does antibiotics treat typhus? Our experiments show just as much of a range of perspectives in answering this question as any from the arts and humanities.

In both cases the quest for truth is influenced by context, which can be infinite in its possibilities and expression. The treatment of typhus, for example, must account for family history, patient history, drug availability, severity, and many more factors. Ultimately, the need to account for and understand this medical context is no different than seeking an account for what makes a good president. Nobody denies the presence of context, but because context is so variable and so difficult to capture, in computer science and AI we often simplify context by treating these base questions as if they had true or false answers.

We can do better. The more recent development of crowdsourcing gives us better access to the range of contextual possibilities. But in order to exploit this and properly capture context we have to let go of our conditioned desire for a single truth and embrace the power of disagreement. Our experiments have shown that, rather than forcing people into a single point of agreement, simply allowing them to disagree by asking the right questions exposes a richer set of possibilities that, with appropriate mathematical frameworks, can finally give us a handle on identifying, processing and understanding context.

CrowdTruth [1–3] is a new approach to understanding semantics, that harnesses the power of the crowd to provide a multitude of perspectives. CrowdTruth uses this scale and diversity to inform a vector space model of truth instead of a boolean, fuzzy, or statistical model. Experimental results show that our approach provides a better way of capturing context, and a more accurate way to predict and explain the way researchers in the sciences and the humanities understand truth.

References

1. Aroyo, L., Welty, C.: The three sides of crowdtruth. Hum. Comput. J. **1**(1) (2014)
2. Aroyo, L., Welty, C.: Truth is a lie: crowd truth and the seven myths of human annotation. AI Mag. **36**(1) (2015)
3. Aroyo, L., Welty, C.: To be AND not to be: quantum intelligence? TEDx talk, Navesink, NJ, April 2015. https://www.youtube.com/watch?v=CyAI_IVUdzM

The Herbrand Manifesto

Thinking Inside the Box

Michael Genesereth and Eric J.Y. Kao

Computer Science Department,
Stanford University
genesereth@stanford.edu
erickao@cs.stanford.edu

Abstract. The traditional semantics for (first-order) relational logic (sometimes called *Tarskian* semantics) is based on the notion of interpretations of constants in terms of objects external to the logic. *Herbrand* semantics is an alternative that is based on truth assignments for ground sentences without reference to external objects. Herbrand semantics is simpler and more intuitive than Tarskian semantics; and, consequently, it is easier to teach and learn.

Moreover, it is more expressive than Tarskian semantics. For example, while it is not possible to finitely axiomatize natural number arithmetic completely with Tarskian semantics, this can be done easily with Herbrand semantics. Herbrand semantics even enables us to define the least fixed-point model of a stratified logic program without any special constructs.

The downside is a loss of some familiar logical properties, such as compactness and proof-theoretic completeness. However, there is no loss of inferential power—anything that can be deduced according to Tarskian semantics can also be deduced according to Herbrand semantics. Based on these results, we argue that there is value in using Herbrand semantics for relational logic in place of Tarskian semantics. It alleviates many of the current problems with relational logic and ultimately may foster a wider use of relational logic in human reasoning and computer applications. To this end, we have already taught several sessions of the computational logic course at Stanford and a popular MOOC using Herbrand semantics, with encouraging results in both cases.

Extending Datalog Intelligence

Benny Kimelfeld

Technion, Israel

Abstract. Prominent sources of Big Data include technological and social trends, such as mobile computing, blogging, and social networking. The means to analyse such data are becoming more accessible with the development of business models like cloud computing, open-source and crowd sourcing. But that data have characteristics that pose challenges to traditional database systems. Due to the uncontrolled nature by which data is produced, much of it is free text, often in informal natural language, leading to computing environments with high levels of uncertainty and error. In this talk I will offer a vision of a database system that aims to facilitate the development of modern data-centric applications, by naturally unifying key functionalities of databases, text analytics, machine learning and artificial intelligence. I will also describe my past research towards pursuing the vision by extensions of *Datalog* — a well studied rule-based programming paradigm that features an inherent integration with the database, and has a robust declarative semantics. These extensions allow for incorporating information extraction from text, and for specifying statistical models by probabilistic programming.

B. Kimelfeld—Taub Fellow – supported by the Taub Foundation.

Analysis and Debugging of Ontology-Based Data Access Specifications

Riccardo Rosati

Dipartimento di Ingegneria informatica,
automatica e gestionale Sapienza Università di Roma

Ontology-based data access (OBDA) is a recent paradigm for accessing *data sources* through an *ontology* that acts as a conceptual, integrated view of the data, and declarative *mappings* that connect the ontology to the data sources. The framework of OBDA has received a lot of attention in the last years: many theoretical studies have paved the way for the construction of OBDA systems and the development of OBDA projects for enterprise data management in various domains.

One important aspect in OBDA concerns the construction of a system specification, i.e., defining the ontology and the mappings over an existing set of data sources. Mappings are indeed the most complex part of an OBDA specification, since they have to capture the semantics of the data sources and express such semantics in terms of the ontology. More precisely, a mapping is a set of assertions, each one associating a query over the source schema with a query over the ontology; the intuitive meaning of a mapping assertion is that all the tuples satisfying the query over the source schema also satisfy the query over the ontology.

The first experiences in the application of the OBDA framework in real-world scenarios have shown that the semantic distance between the conceptual and the data layer is often very large, because data sources are mostly application-oriented: this makes the definition, debugging, and maintenance of mappings a hard and complex task. Such experiences have clearly shown the need of tools for supporting the management of mappings. However, so far no specific approach has explicitly dealt with the problem of mapping analysis and evolution in the context of OBDA. The work on schema mappings in data exchange, probably the closest one to mapping management in OBDA, has considered the problem of analyzing the formal properties of mappings, but in a different framework and under different assumptions on the schema languages.

In this talk, we will present some recent results on mapping analysis and evolution in OBDA obtained in the context of the Optique European project[1]. More precisely, we will first introduce basic notions of mapping inconsistency and mapping redundancy in an OBDA specification. Then, based on such notions, we will present a computational analysis of the problem of checking the above anomalies in an OBDA specification, for a wide range of ontology languages and for different mapping languages. Finally, we will focus on mapping evolution in OBDA, providing formal definitions and computational results for the basic forms of mapping update when the ontology is changed.

---

[1] www.optique-project.eu.

This is joint work with Domenico Lembo, José Mora, Domenico Fabio Savo, and Evgenij Thorstensen. This research has been partially supported by the EU under FP7 project Optique (grant n. FP7-318338).

Contents

Extending Datalog Intelligence

Benny Kimelfeld[(✉)]

Technion, Haifa, Israel
bennyk@cs.technion.ac.il

Abstract. Prominent sources of Big Data include technological and social trends, such as mobile computing, blogging, and social networking. The means to analyse such data are becoming more accessible with the development of business models like cloud computing, open-source and crowd sourcing. But that data have characteristics that pose challenges to traditional database systems. Due to the uncontrolled nature by which data is produced, much of it is free text, often in informal natural language, leading to computing environments with high levels of uncertainty and error. In this talk I will offer a vision of a database system that aims to facilitate the development of modern data-centric applications, by naturally unifying key functionalities of databases, text analytics, machine learning and artificial intelligence. I will also describe my past research towards pursuing the vision by extensions of *Datalog* — a well studied rule-based programming paradigm that features an inherent integration with the database, and has a robust declarative semantics. These extensions allow for incorporating information extraction from text, and for specifying statistical models by probabilistic programming.

1 Introduction

The management and analysis of massive data volumes has been practiced in the past decades by large organizations such as enterprises and governmental agencies. In the Big Data era, massive volumes of data have become accessible to the larger public. Technological and social trends, such as *mobile computing*, *blogging* and *social networking*, result in data with a high potential value for a plethora of domains, such as business intelligence, marketing, civil services, and political movements. Moreover, contemporary business models, such as *cloud computing*, *open source* and *crowd sourcing*, provide the means to analyse such data without requiring the vast resources of grand enterprises. But that data have characteristics that introduce new challenges to database management systems. The uncontrolled nature by which data is generated (e.g., people casually post statements via mobile devices) implies that much of the data is free text in informal natural language, where standard processors involve non-negligible levels of error and uncertainty. The historical success of general-purpose database systems is largely due to the access and query model, most notably SQL that allows to phrase queries in a simple and intuitive language and, hence, facilitates data management for a large community of developers. But these systems

© Springer International Publishing Switzerland 2015
B. ten Cate and A. Mileo (Eds.): RR 2015, LNCS 9209, pp. 1–10, 2015.
DOI: 10.1007/978-3-319-22002-4_1

fall short of providing the suitable means for nontrivial extraction of information from text, and therefore, even basic operations in the analysis of Big Data instances require integration with out-of-database paradigms.

More specifically, applications that involve text analytics often bundle together functions for extracting text fragments from documents (usually by means of scripting languages), statistical or machine-learning libraries to filter out errors, and a relational database to incorporate the extracted data with structured data [43,44]. The application of these different paradigms requires different skills and development styles (e.g., script coding, database querying, and machine-learning engineering), and is often carried out by different (groups of) developers. This practice entails a complicated and laborious development. For example, joining extracted data with structured data may reveal low recall (coverage) in the extracted data, which may require the extraction phase to produce more data; so the extraction developers expand the extraction, but then this expansion is found by the statistics developers to be overly costly for their computation. This problem may often be solved if one knows to begin with what the extracted data will later join with (and, hence, avoid irrelevant extracted output); so the extraction developers are either importing structured data from the database developers, or decide to integrate with the database before the statistical phase. Similarly, when cleaning the extracted data via machine learning, feature engineering (commonly considered as the most challenging part in developing a machine-learning solution [29,61]) requires repeated communication between the paradigms.

There are additional drawbacks to programming over separate paradigms and platforms. One is the loss of opportunities for automatic optimization that involve cross-component insights, as illustrated in past research [49,53]. For example, the above early join (or filtering) could very well be carried out automatically by a query planner, rather then the developer. Another drawback is in limiting the ability to establish precise confidence estimation that requires a holistic view of the execution. For example, it may be the case that a large number of answers indicate some uncertain information (e.g., the sentiment of a post), but they are all based on one early decision that itself has a low confidence.

This talk will describe a vision of a database system that naturally integrates text extraction within a relational database system. The vision involves three design principles that we believe are essential for such an integration: a *unified model*, *underspecification* (i.e., the ability to allow machine learning to complement query specification), and a *probabilistic interpretation*. The vision is based on extending *Datalog*, and the talk will describe our past research towards that. Datalog is a *purely declarative* query (or database programming) language in a strong sense: it constitutes a set of rules with semantics that is independent of any execution order, and it is invariant under logical equivalence (i.e., if a rule is already implied by others, then its inclusion/exclusion does not change the program). Moreover, its inherent support of *recursion* allows to phrase complicated programs quite easily and compactly. Datalog has been extensively studied by the database-research community, and has several commercial instantiations, such as LogicBlox [26].

2   Frameworks for Information Extraction

The core operation required for querying textual data is that of *Information Extraction* (IE). The goal in IE is to populate a predefined relational schema that has predetermined underlying semantics, by correctly detecting the values of records in a given text document (or a collection thereof). Popular tasks in the space of IE include *named entity recognition* [54] (identify proper names in text, and classify those into a predefined set of categories such as *person* and *organization*), *relation extraction* [59] (extract tuples of entities that satisfy a predefined relationship, such as *person-organization*), *event extraction* [2] (find events of predefined types along with their key players, such as *nomination* and *nominee*), *temporal information extraction* [20,37] (associate mentions of facts with mentions of their validity period, such as *nomination-date*), and *coreference resolution* [48] (match between phrases that refer to the same entity, such as "Obama," "the President," and "him").

Significant efforts have been put to design and establish development frameworks for IE. Xlog [53] extends Datalog with special primitive types such as documents (distinguished chunks of text) and spans (intervals of text within a document), and matchers of regular expressions. One of the most commonly used IE systems is the General Architecture for Text Engineering (GATE) [19], an open-source project by the University of Sheffield, which is an instantiation of the *cascaded finite-state transducers* [3]. In GATE, a document is processed by a sequence of phases (cascades), each annotating spans with types by applying grammar rules over previous annotations. SystemT [15] is IBM's principal IE tool that features an SQL-like declarative language named AQL (Annotation Query Language), along with a query-plan optimizer [49] and development tooling [38]. This system is typically integrated within a larger software bundle for data analytics. Other industrial data analytics tools that support IE development include Attensity, Clarabridge, IBM BigInsights, HP Autonomy, Oracle Collective Intellect, SAS, SAP, and more. Stanford's DeepDive [44] supports a language to combine various technologies, such as Python scripts for preliminary IE, CSV readers, a PostgreSQL database, and an inference engine for Markov-Logic Networks [47,50].

Development frameworks like XLog, GATE and SystemT provide different mechanisms and languages to query text by means of rules. The underlying assumption is that a solution for the IE task at hand can be phrased as a collection of operations that are deterministic in nature, fully specified, and manually encoded. Often, however, the IE task and the underlying textual data have inherent properties that violate this assumption. In particular, traditional rules can hardly accommodate the flexibility by which natural language can be used to express knowledge, alongside common human practices like informality, deceit and sarcasm. The gap between this traditional programming philosophy and the nature of the text of interest is typically managed by long chains of rules, filters over rules, exceptions over filters, and so on. In contrast, the Natural Language Processing (NLP) community has been focusing on statistical-modelling techniques far more than rules-based approaches. Examples include naïve Bayes classifiers, and various kinds of probabilistic graphical models such as hidden Markov models [9,10,24,36], maximum entropy Markov models [33,40], and Conditional Random Fields (CRF) [14,35,58], that model the text and the annotation as a probabilistic generative model with specified dependencies (edges). A recent study by IBM researchers [16] highlights and quantifies this discrepancy between the disregard of rule-based approaches in scientific publications on IE, and their dominance in industrial solutions.

However, a toolkit of algorithmic techniques is far from being sufficient for a general-purpose development framework, and in fact, systems such as GATE, SystemT and Xlog have good reasons to prefer rules to the aforementioned statistical approaches. Significant skills in Computer Science are required to deploy, adapt and scale up these techniques to the specific use case of interest. Moreover, business considerations may require high flexibility in the behaviour of extractors, such as avoidance of specific types of mistakes; such flexibility is inherently built in the rule-based approaches, while it is often a challenge to tweak a statistical solution, or generally to express different kinds of domain knowledge. Gupta and Manning [27] explain that *"[. . .] rules are effective, interpretable, and are easy to customize by non-experts to cope with errors."* Towards closing this gap, some data management systems (e.g., MADden [25]) have been designed with specific built-in extractors (e.g., part-of-speech taggers, sentiment detectors, and named-entity recognizers) that are implemented using statistical models; these extractors aim to capture common IE needs, but are not designed to enable developers to build special extractors for their own use case (e.g., finding specific chemical interactions mentioned in scientific manuscripts).

3   Design Principles

We envision a general-purpose database system that captures the nature of text-centric data, in order to facilitate, expedite, and simplify the development of applications thereof by a wide range of developers. Towards that, we propose to fundamentally revise the basic models of *database* and *querying*. Instead of a collection of rules that encode fully specified logical assertions, a query is a combination of rigid rules, soft rules (or features), and unspecified rules (which are placeholders for subqueries learned from training examples), with an inherent access to text and auxiliary NLP machinery. In order to balance between precision and recall, the developer does not need to redesign the (typically complex) query, but rather to adjust the *probability* by which answers are required to be correct. From this goal we distill three fundamental design principles.

- **Unified Model.** This principle means that we pursue a data and query model that will allow to easily query structured data and textual data in a unified and elegant manner. In terms of the data model, this entails more than just *storing* both types of data, but rather representing the intermediate information that is crucial for text analysis (e.g., where in the text the extraction took place). In terms of the query model, such a unification should allow for subqueries that can join previously extracted data, generic NLP analysis, and structured data.

- **Underspecification.** This principle implies that the model should *utilize* techniques of machine learning and artificial intelligence, rather than *replace* them. We pursue a design that will allow to control the level of automation, and to effectively integrate it with the manual rules. In particular, these techniques will not necessarily take over the entire program, but will rather take over subqueries that the developer believes can be more effectively pursued by automated learning techniques. Moreover, the level of underspecification should be flexible: the user may provide a complete set of features in one task, and ask for automatic feature generation in another.
- **Probabilistic Interpretation.** This principle means that, conceptually, a program produces not just answers to a query, but rather a probability space of such answers. The probability space is defined using the precision of the learning components, as well as the confidence levels provided by the generic NLP executions.

Next, we discuss each of the three design principles in more detail.

3.1 Unified Model

We plan to pursue a unified data and query model by building on the recently developed concept of *document spanners* (or simply *spanners* for short) [22,23,30], which is inspired by SystemT [15]. In that formalism, primitive extractors construct base relations over spans from the text, while a relational query language (e.g., relational algebra or Datalog) manipulates these relations. A common example of primitive extractors are regular expression with *capture variables*, which we call a *regex-formula*. At the conceptual level, a spanner is simply a function that maps a given document (string) into a relation, of a predefined schema, over the document's *spans* (that is, document intervals that are identified by their beginning and ending index).

The spanner framework features a clean theoretical model, giving rise to a nontrivial investigation via corresponding types of *transducers*. In particular, we have defined simple extensions of nondeterministic (finite-state) automata that are able to assign spans to variables as they run. We have further identified such automata that capture precisely the closure of regex formulas under the relational algebra, and under non-recursive Datalog.

The equivalence between the Datalog and the automata representation has given rise to several fundamental results. For instance, it allowed us to explore the implication of key extensions to the language, such as the complementation operator and the string-equality predicate. Moreover, it allowed to draw strong connections to past literature, such as *string relations* [7] and *graph queries* [8]. Finally, it was the basis of a more recent work [22] where we have shown that the ad-hoc *cleaning strategies* in rule-based IE systems can be cast as instances of the well studied concept of *database repairs* [4] in the presence of tuple priorities [55], giving rise to fundamental analysis such as *well definedness* and *expressive power*. The framework of prioritized repairs is fairly recent, and not much is known about fundamental aspects such as the computational complexity it entails.

In a recent work [21] we have studied that complexity in a traditional database setting (that does not involve text analysis).

Our plan is to generalize the relational data model with *span attributes* that reference strings in the database. This extension will allow us to extend Datalog with spanners, and hence, establish a unified language to query structured and textual data simultaneously and interchangeably. In traditional Datalog terminology, we will extend the *extensional database* (i.e., the supplied relations without the inferred relations) with spanners, and apply traditional inference thereof. Spanners will come from various sources, including developer-phrased regex formulas, and built-in algorithms for NLP. We note that such algorithms may involve *grammar correction* (or *normalization*) to enable the usability of linguistic parsers [60].

3.2 Underspecification

We identify two extremes of underspecified queries. The "easier" extreme is where the developer specifies the parts of the data (textual or relational) that should be considered to infer the desired relation, but she does not provide the precise way of combining those parts into a query that carries out the actual inference. Instead, she provides (positive and negative) examples of the desired result. This is the vanilla setting in *supervised machine learning*, where those parts are referred to as *features*. A learning algorithm is associated with a class of functions (e.g., Datalog rules, decision trees, linear classifiers, and graphical models), called *models*, and it aims to find in that class a model that fits the examples, and more importantly, well generalizes to unseen examples. Naturally, the model class has a crucial impact on the quality of the result, as well as on the complexity of learning the model and evaluating it thereafter.

The "harder" extreme is where the developer provides *only* examples, and no features. The system is then expected to use all of its available knowledge in order to come up with both the features and model. We believe that this scenario is quite common, as feature engineering is commonly regarded highly challenging and laborious [29,61]. Fortunately, there has been a substantial research on techniques to infer functions in the absence of features. These include *rule induction* (or *inductive logic programming*) [17,54], graph-kernel methods [59,62], *structure learning* in statistical relational learning [34,41], and *frequent subgraph mining* [13,39,45,51]. In a recent work [31], we have conducted a thorough study of the complexity of frequent subgraph mining, and proposed a novel algorithmic approach (based on the notion of *hereditary graph properties* [18]) that provides complexity guarantees that were are not featured in previous algorithms (e.g., [28,57]). We plan to investigate the translation of this novel approach into an effective implementation.

3.3 Probabilisitic Interpretation

While ordinary Datalog can be thought of as a deterministic function that maps a given input database into an output database (or *outcome*), a probabilistic variant of Datalog maps the input database into a *probability space* over possible outcomes. Here again there is a vast literature on such variants. These variants are typically logical specifications of graphical models—directed models [5,42] and undirected models [11,50].

One drawback of applying existing notions of probabilistic logic to Datalog is that we are forced to compromise important benefits of the ordinary Datalog. The notions that are driven by directed graphical models require an *order* of rule execution, and hence, impose programs that are sensitive to the execution order (in constrast to being only a *set* of rules). The models that are driven by the undirected graphical models are sensitive to the specifics (e.g., number) of grounding of each rule. This means, for instance, that adding a rule that is equivalent to another rule may have a substantial impact on the semantics of the program. Hence, such models violate invariance under logical equivalence.

A class of probabilistic logical models that retains the pure declarative nature of Datalog includes [1,32,46,52]. These models are based on a separation between the probabilistic choices and the rule specification, in the spirit of *probabilistic databases* [56]; that is, execution is conceptually a two-step process: in the first step a random database is bring generated by an external process, and in the second phase logic is applied. In a recent work [6], we have proposed *Probabilistic Programming Datalog (PPDL)*, which is a framework that extends Datalog with convenient mechanisms to include common numerical probability functions. In particular, conclusions of rules may contain values drawn from such functions. The framework is based on *tuple-generating dependencies* with existential variables [12], and it unifies the concepts of database *integrity constraints* and *statistical observations*. PPDL is invariant under logical equivalence (by its definition), and our analysis has shown that it is invariant under execution (chase) order.

4   Concluding Remarks

We described a vision of a system that elegantly incorporates text analysis within structured-data management. We identified three design principles that we believe are essential for such a system: *unified model, underspecification,* and *probabilistic interpretation.* For each of these principles, we described our relevant past research: document spanners, frequent subgraph mining, and PPDL, respectively. By applying our past research and the vast knowledge acquired by other systems (e.g., SystemT [15] and DeepDive [44]), our next steps will be to establish a full system model, design, and implementation.

References

1. Abiteboul, S., Deutch, D., Vianu, V.: Deduction with contradictions in Datalog. In: ICDT, pp. 143–154 (2014)
2. Aone, C., Ramos-Santacruz, M.: Rees: a large-scale relation and event extraction system. In: ANLP, pp. 76–83 (2000)

3. Appelt, D.E., Onyshkevych, B.: The common pattern specification language. In: Proceedings of the TIPSTER Text Program: Phase III, pp. 23–30, Baltimore, Maryland, USA (1998)
4. Arenas, M., Bertossi, L.E., Chomicki, J.: Consistent query answers in inconsistent databases. In: PODS, pp. 68–79 (1999)
5. Baral, C., Gelfond, M., Rushton, N.: Probabilistic reasoning with answer sets. Theory Pract. Log. Program. **9**(1), 57–144 (2009)
6. Barany, V., Cate, B.T., Kimelfeld, B., Olteanu, D., Vagena, Z.: Declarative statistical modeling with datalog (2014). arXiv preprint arXiv:1412.2221
7. Barceló, P., Figueira, D., Libkin, L.: Graph logics with rational relations and the generalized intersection problem. In: LICS, pp. 115–124 (2012)
8. Barceló, P., Libkin, L., Lin, A.W., Wood, P.T.: Expressive languages for path queries over graph-structured data. ACM Trans. Database Syst. **37**(4), 31 (2012)
9. Bikel, D.M., Miller, S., Schwartz, R.M., Weischedel, R.M.: Nymble: a high-performance learning name-finder. In: ANLP, pp. 194–201 (1997)
10. Borkar, V.R., Deshmukh, K., Sarawagi, S.: Automatic segmentation of text into structured records. In: SIGMOD Conference, pp. 175–186. ACM (2001)
11. Bröcheler, M., Mihalkova, L., Getoor, L.: Probabilistic similarity logic. In: UAI, pp. 73–82. AUAI Press (2010)
12. Calì, A., Gottlob, G., Lukasiewicz, T., Marnette, B., Pieris, A.: Datalog+/-: a family of logical knowledge representation and query languages for new applications. In: LICS, pp. 228–242 (2010)
13. Chakravarthy, S., Venkatachalam, A., Telang, A., Aery, M.: Infosift: a novel, mining-based framework for document classification. IJNGC 5(2) (2014)
14. Chen, F., Feng, X., Re, C., Wang, M.: Optimizing statistical information extraction programs over evolving text. In: ICDE, pp. 870–881. IEEE Computer Society (2012)
15. Chiticariu, L., Krishnamurthy, R., Li, Y., Raghavan, S., Reiss, F., Vaithyanathan, S.: SystemT: an algebraic approach to declarative information extraction. In: ACL, pp. 128–137 (2010)
16. Chiticariu, L., Li, Y., Reiss, F.R.: Rule-based information extraction is dead! Long live rule-based information extraction systems! In: EMNLP, pp. 827–832. ACL (2013)
17. Ciravegna, F.: Adaptive information extraction from text by rule induction and generalisation. In: IJCAI, pp. 1251–1256. Morgan Kaufmann (2001)
18. Cohen, S., Kimelfeld, B., Sagiv, Y.: Generating all maximal induced subgraphs for hereditary and connected-hereditary graph properties. J. Comput. Syst. Sci. **74**(7), 1147–1159 (2008)
19. Cunningham, H.: GATE: a general architecture for text engineering. Comput. Humanit. **36**(2), 223–254 (2002)
20. Dylla, M., Miliaraki, I., Theobald, M.: A temporal-probabilistic database model for information extraction. PVLDB **6**(14), 1810–1821 (2013)
21. Fagin, R., Kimelfeld, B., Kolaitis, P.G.: Dichotomies in the complexity of preferred repairs. In: PODS 2015 (2015) (To appear)
22. Fagin, R., Kimelfeld, B., Reiss, F., Vansummeren, S.: Cleaning inconsistencies in information extraction via prioritized repairs. In: PODS. ACM (2014)
23. Fagin, R., Kimelfeld, B., Reiss, F., Vansummeren, S.: Document spanners: a formal approach to information extraction. J. ACM (JACM) **62**(2), 12 (2015)
24. Ginsburg, S., Wang, X.S.: Regular sequence operations and their use in database queries. J. Comput. Syst. Sci. **56**(1), 1–26 (1998)

25. Grant, C.E., Gumbs, J., Li, K., Wang, D.Z., Chitouras, G.: Madden: query-driven statistical text analytics. In: CIKM, pp. 2740–2742. ACM (2012)
26. Green, T.J., Aref, M., Karvounarakis, G.: LogicBlox, platform and language: a tutorial. In: Barceló, P., Pichler, R. (eds.) Datalog 2.0 2012. LNCS, vol. 7494, pp. 1–8. Springer, Heidelberg (2012)
27. Gupta, S., Manning, C.D.: Improved pattern learning for bootstrapped entity extraction. In: CoNLL, pp. 98–108. ACL (2014)
28. Huan, J., Wang, W., Prins, J., Yang, J.: SPIN: mining maximal frequent subgraphs from graph databases. In: KDD, pp. 581–586 (2004)
29. Kandel, S., Paepcke, A., Hellerstein, J.M., Heer, J.: Enterprise data analysis and visualization: an interview study. IEEE Trans. Vis. Comput. Graph. **18**(12), 2917–2926 (2012)
30. Kimelfeld, B.: Database principles in information extraction. In: PODS, pp. 156–163. ACM (2014)
31. Kimelfeld, B., Kolaitis, P.G.: The complexity of mining maximal frequent subgraphs. ACM Trans. Database Syst. **39**(4), 32:1–32:33 (2014)
32. Kimmig, A., Demoen, B., De Raedt, L., Santos Costa, V., Rocha, R.: On the implementation of the probabilistic logic programming language ProbLog. Theory Pract. Logic Program. **11**, 235–262 (2011)
33. Klein, D., Manning, C.D.: Conditional structure versus conditional estimation in NLP models. In: EMNLP, pp. 9–16. Association for Computational Linguistics (2002)
34. Kok, S., Domingos, P.M.: Learning markov logic networks using structural motifs. In: ICML, pp. 551–558. Omnipress (2010)
35. Lafferty, J.D., McCallum, A., Pereira, F.C.N.: Conditional random fields: probabilistic models for segmenting and labeling sequence data. In: ICML, pp. 282–289 (2001)
36. Leek, T.R.: Information extraction using hidden Markov models. Master's thesis, UC San Diego (1997)
37. Ling, X., Weld, D.S.: Temporal information extraction. In AAAI. AAAI Press (2010)
38. Liu, B., Chiticariu, L., Chu, V., Jagadish, H.V., Reiss, F.: Automatic rule refinement for information extraction. PVLDB **3**(1), 588–597 (2010)
39. Matsumoto, S., Takamura, H., Okumura, M.: Sentiment classification using word sub-sequences and dependency sub-trees. In: Ho, T.-B., Cheung, D., Liu, H. (eds.) PAKDD 2005. LNCS (LNAI), vol. 3518, pp. 301–311. Springer, Heidelberg (2005)
40. McCallum, A., Freitag, D., Pereira, F.C.N.: Maximum entropy Markov models for information extraction and segmentation. In: ICML, pp. 591–598 (2000)
41. Mihalkova, L., Mooney, R.J.: Bottom-up learning of Markov logic network structure. In: ICML, pp. 625–632. ACM (2007)
42. Milch, B., et al: BLOG: probabilistic models with unknown objects. In: IJCAI, pp. 1352–1359 (2005)
43. Niu, F., Ré, C., Doan, A., Shavlik, J.W.: Tuffy: scaling up statistical inference in Markov logic networks using an RDBMS. PVLDB **4**(6), 373–384 (2011)
44. Niu, F., Zhang, C., Re, C., Shavlik, J.W.: DeepDive: Web-scale knowledge-base construction using statistical learning and inference. In: Proceedings of the Second International Workshop on Searching and Integrating New Web Data Sources, CEUR Workshop Proceedings, vol. 884, pp. 25–28 (2012). http://CEUR-WS.org
45. Pons-Porrata, A., Llavori, R.B., Ruiz-Shulcloper, J.: Topic discovery based on text mining techniques. Inf. Process. Manage. **43**(3), 752–768 (2007)

46. Poole, D.: The independent choice logic and beyond. In: De Raedt, L., Frasconi, P., Kersting, K., Muggleton, S.H. (eds.) Probabilistic Inductive Logic Programming. LNCS (LNAI), vol. 4911, pp. 222–243. Springer, Heidelberg (2008)

47. Poon, H., Domingos, P.: Joint inference in information extraction. In: Proceedings of the 22nd national conference on Artificial intelligence, AAAI 2007, pp. 913–918. AAAI Press (2007)

48. Raghunathan, K., Lee, H., Rangarajan, S., Chambers, N., Surdeanu, M., Jurafsky, D., Manning, C.D.: A multi-pass sieve for coreference resolution. In: EMNLP, pp. 492–501. ACL (2010)

49. Reiss, F., Raghavan, S., Krishnamurthy, R., Zhu, H., Vaithyanathan, S.: An algebraic approach to rule-based information extraction. In: ICDE, pp. 933–942 (2008)

50. Richardson, M., Domingos, P.: Markov logic networks. Mach. Learn. **62**(1–2), 107–136 (2006)

51. Rink, B., Bejan, C.A., Harabagiu, S.M.: Learning textual graph patterns to detect causal event relations. In: Proceedings of the Twenty-Third International Florida Artificial Intelligence Research Society Conference. AAAI Press (2010)

52. Sato, T., Kameya, Y.: PRISM: a language for symbolic-statistical modeling. In: IJCAI, pp. 1330–1339 (1997)

53. Shen, W., Doan, A., Naughton, J.F., Ramakrishnan, R.: Declarative information extraction using datalog with embedded extraction predicates. In: VLDB, pp. 1033–1044 (2007)

54. Soderland, S.: Learning information extraction rules for semi-structured and free text. Mach. Learn. **34**(1–3), 233–272 (1999)

55. Staworko, S., Chomicki, J., Marcinkowski, J.: Prioritized repairing and consistent query answering in relational databases. Ann. Math. Artif. Intell. **64**(2–3), 209–246 (2012)

56. Suciu, D., Olteanu, D., Ré, C., Koch, C.: Probabilistic Databases. Synthesis Lectures on Data Management. Morgan & Claypool Publishers, San Rafael (2011)

57. Thomas, L.T., Valluri, S.R., Karlapalem, K.: Margin: Maximal frequent subgraph mining. TKDD 4(3) (2010)

58. Wang, D.Z., Franklin, M.J., Garofalakis, M.N., Hellerstein, J.M., Wick, M.L.: Hybrid in-database inference for declarative information extraction. In: SIGMOD Conference, pp. 517–528. ACM (2011)

59. Zelenko, D., Aone, C., Richardella, A.: Kernel methods for relation extraction. J. Mach. Learn. Res. **3**, 1083–1106 (2003)

60. Zhang, C., Baldwin, T., Ho, H., Kimelfeld, B., Li, Y.: Adaptive parser-centric text normalization. In: ACL, vol. 1, pp. 1159–1168. The Association for Computer Linguistics (2013)

61. Zhang, C., Kumar, A., Ré, C.: Materialization optimizations for feature selection workloads. In: SIGMOD Conference, pp. 265–276 (2014)

62. Zhao, S., Grishman, R.: Extracting relations with integrated information using kernel methods. In ACL. The Association for Computer Linguistics (2005)

An Ontology for Historical Research Documents

Giovanni Adorni[1], Marco Maratea[1], Laura Pandolfo[1], and Luca Pulina[2(✉)]

[1] DIBRIS, Università di Genova, Via Opera Pia, 13, 16145 Genova, Italy
{adorni,marco.maratea}@unige.it, laura.pandolfo@edu.unige.it
[2] POLCOMING, Università di Sassari, Viale Mancini N. 5, 07100 Sassari, Italy
lpulina@uniss.it

Abstract. In this paper we present the conceptual layer of STOLE, our ontology-based digital archive aiming at helping historical researchers to organize data, extract information and derive new knowledge from historical documents.

1 Context and Motivation

Historical documents are considered a rich and valuable source of information related to, e.g., events and people used by researchers and scholars to investigate history. In the last decades, the digitization of historical documents has been mainly focused on developing applications that enable users to access, retrieve and query information in a highly efficient way [1]. In fact, historical documents are characterized by being syntactically and semantically heterogeneous, semantically rich, multilingual, and highly interlinked. They are usually produced in a distributed, open fashion by organizations like museums, libraries, and archives, using their own established standards and best practices [2].

It is well-established that ontologies can offer a clear conceptual representation and they provide a valuable support to knowledge extraction, knowledge discovering, and data integration. More, they can offer effective solutions about design and implementation of user-friendly ways to access and query content and meta-data – see [3] for a survey in the historical research domain.

In this paper we describe the STOLE[1] ontology, that represents the conceptual layer of our ontology-based digital archive. The main goal of the STOLE ontology is to clearly model historical concepts and, at the same time, to gain insights into this specific field, e.g., supporting historians to find out some unexplored but useful aspects about a particular event or person. STOLE collects information about some of the most relevant journal articles published between 1848 and 1946 concerning the legislative history of public administration in Italy. These documents are regarded as an estimable source of information for historical research since through the study of these texts it is possible to trace the course of Italian history.

---

[1] STOLE is the acronym for the Italian "STOria LEgislativa della pubblica amministrazione italiana", that means "Legislative History of Italian Public Administration".

© Springer International Publishing Switzerland 2015
B. ten Cate and A. Mileo (Eds.): RR 2015, LNCS 9209, pp. 11–18, 2015.
DOI: 10.1007/978-3-319-22002-4_2

The rest of the paper is organized as follows: in Sect. 2 we provide some explanations about the main steps and the key decisions which marked the ontology design process, and we describe the STOLE ontology in detail. In Sect. 3 we briefly describe the architecture of our ontology-based digital archive in order to provide an overall view of the system. Finally, we conclude the paper in Sect. 4 with some final remarks and future work.

2   The STOLE Ontology

2.1   Design At-a-Glance

Narrative and statistical documents represent the main sources used by historians to conduct their research. Typically, historians want to extract the facts from these documents in order to gather information for reconstructing specific historical events.

Our design process derived from the needs of the researchers of Department of History of the University of Sassari which, since the 1980s, have been involved in a project designed to collect, digitalize and catalogue historical journals concerning genesis and evolution of the Italian public administrations and institutions. The research was conducted on a wide selection of magazines owned by the following Italian institutions' libraries: Central Archives of the State, Chamber of Deputies, Supreme Court of Cassation, University of Bologna and University of Sassari. As a result, the ARAP[2] archive of the University of Sassari was created and it actually collects a large amount of narrative sources. Currently, historians can access to several websites which allow them to flip between pages of relevant documents, however effective analysis tools are rarely provided by these applications. Semantic web technologies can address some specific issues in historical domain, since they allow to identify implicitly and explicitly knowledge included in the documents. For example, reference to a historical person contained in a historical source can be discovered and related to other entities, e.g., events in which that person participated, providing a rich representation from which historians can extract meaningful knowledge to their research [3].

In the following, we can summarize the main phases of the creation process:

1. Identification of key concepts.
2. Identification of the proper language and Tbox implementation.
3. Ontology population, i.e., filling the Abox with semantic annotations.

In the first step the domain experts have been involved in order to contribute to the definition of key issues related to the application domain. In particular, we detected the main categories of data expressed in the considered historical documents. The results of this process enabled us to compute a taxonomy composed of the following three elements:

---

[2] ARAP is the acronym for the Italian "Archivio di Riviste sull'Amministrazione Pubblica", that means "Archive of Journals on Italian Public Administration".

Table 1. Tbox statistics about the STOLE ontology. These data are computed by PROTÉGÉ [5] in the *Metrics* view.

Classes	14
Axioms	440
Object properties	30
Data properties	29

- Data concerning the author of the article, e.g., name, surname and biography.
- Data concerning the journal and the article, e.g., article title, journal name, date and topics raised in the article.
- Data concerning some relevant facts and persons cited in the article, e.g., persons, historical events, institutions.

Historical analysis in this specific domain is based on the above information and focused on the interrelations between these data. For example, the link between an author and the people cited in an article provides valuable information to historians, e.g., if an author has often referred to King Vittorio Emanuele II probably it can easily be interpreted as favorable to the monarchy.

Regarding the second point, the Tbox of the STOLE ontology has been designed building on some existing standards and meta-data vocabularies, such as Dublin Core (http://dublincore.org), FOAF (http://www.foaf-project. org), the Bio Vocabulary (http://vocab.org/bio/0.1), the Bibliographic Ontology (http://bibliontology.com), and the Ontology of the Chamber of Deputies (http://dati.camera.it/data/en). In particular, the latter is an ontology aiming at modeling the domain of the Chamber during its history. It can be a relevant source since, for example, most part of the authors in our archive were also involved in government activities.

Concerning the modeling language, our choice fall to OWL2 DL [4]. This language allows us to have proper expressivity, and to model the knowledge for our application by means of constructs like cardinality restrictions and other role constraints, e.g., functional properties.

2.2 Implementation

In the following, we describe main classes, object properties and data properties of our ontology. Statistics are summarized in Table 1.

Article represents our library, namely the collection of historical journal articles. Every instance of this class has data properties such as articleTitle, articleDate, pageStart, and pageEnd.

Jurisprudence is a subclass of Article and contains a series of verdicts which are entirely written in the articles. Every individual of this subclass has the following data properties: sentenceDate, sentenceTitle, and byCourt.

Law is also a subclass of `Article`, and it contains a set of principles, rules, and regulations set up by a government or other authority which are entirely written in the articles. This subclass has data properties such as `lawDate` and `lawTitle`.

Event denotes relevant events. It contains five subclasses modeling different kinds of events: `Birth` and `Death` are subclasses related to a person's life; `BeginPublication` and `EndPublication` represent the publication period of a journal; `HistoricalEvent` contains the most relevant events that have marked the Italian history.

Journal denotes the collection of historical journals. This class has data properties such as `journalArticle`, `publisher`, and `issn`.

Person is the class representing people involved in the Italian legislative and public administration history. This class contains one subclass, `Author`, that includes the contributors of the articles. Every instance of this class has some data properties as `firstName`, `surname`, and `biography`.

Place represents cities and countries related to people and events.

Subject is a class representing topics tackled in the historical journals.

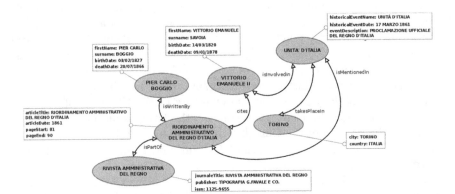

Fig. 1. Example of individuals and their relationships in the STOLE ontology. Ellipses denotes individuals, while information related to their data properties are reported in boxes. Object properties (and their inverse) are denoted by arrows.

Concerning object properties, we describe in the following the ones related to `HistoricalEvent`. The full documentation of the STOLE ontology is available at http://visionlab.uniss.it/STOLE_DOC. `HistoricalEvent` has a crucial role in our ontology, and their relationships with other classes are extremely useful for historical research in order to highlight connections between events, people, and articles. In Fig. 1 we show a graphical example of these relations. Looking at the figure, we can see that Unità d'Italia is an instance of `HistoricalEvent`, and represents one of the most important historical event in Italian history. This event is in relation to individuals in `Article` by the `isMentionedIn` property: this relationship defines in which articles is mentioned the event Unità d'Italia. The `hasWritten` object property relates an article to its author, and `isPartOf` shows in which journal has been published that specific article. With reference to `HistoricalEvent` class, there are further interesting relationships to emphasize:

- the `takesPlaceIn` object property makes explicit the event's location – in the example depicted in Fig. 1 is the city of Torino;
- `InvolvedIn` connects individuals in `Person` class to a particular event, taking into account all people that played a role in a given historical event.

Summing up, the example shown in Fig. 1 hints at potentially interesting relationships among elements that can be represented by the STOLE ontology. Our ontology-based application – that will be described in the next section – supports historians in their research, providing to them relations between events, people and documents in an automated way.

Currently, there are no other examples of ontologies that model this particular domain. However, despite its specific nature, STOLE ontology can be used in different application field that relates the history of the Italian administrations and institutions.

Finally, the ontology has been populated leveraging a set of annotated historical documents comprised into the ARAP archive. Semantic annotations were provided by a team of domain experts and individuals were added to the ontology by means of a JAVA program built on top of the OWL APIs [6].

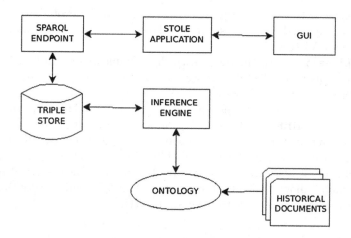

Fig. 2. The architecture of STOLE.

3 System Architecture

In Fig. 2 we report the architecture of our ontology-based digital archive built on top of the conceptual layer represented by the STOLE ontology. Looking at the figure, we can see that it is composed of the modules listed in the following:

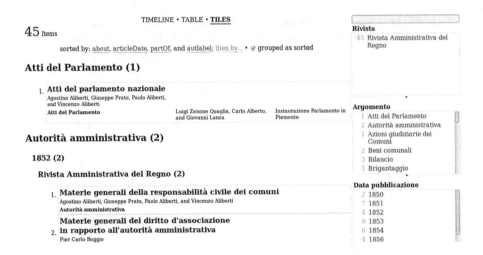

Fig. 3. Screenshot of the STOLE web GUI.

Ontology is the ontology described in Sect. 2.2.

Inference Engine aims to accomplish both classification and consistency checking tasks on the STOLE ontology. It interacts with the Ontology in order to
infer new knowledge to present to the user. Actually, we are using the HermiT reasoner [7].

Triple Store and SPARQL Endpoint are the modules devoted to store and
query the knowledge base, respectively. For these purposes, we are currently
using Open Virtuoso[3].

STOLE Application is the module in which we implemented all functionalities
related to query the SPARQL Endpoint and to process the answer in order
to be presented to the user by means of the GUI.

GUI is devoted to the user-system interaction. This module is implemented on
top of the SIMILE Exhibit API [8], a set of JavaScript files that allows
to easily create rich interactive web pages including maps, timelines, and
galleries with very detailed client-side filtering. This kind of representations,
e.g. timeline of historical events, are widely used in the historical research
field. Exhibit allows to display the result of SPARQL queries in JSON format.
Figure 3 shows a screen-shot of the STOLE GUI.

4   Conclusions and Future Work

In this paper we described the development and implementation of an ontology for historical research documents, and we presented a general architecture
overview about the related ontology-based archive.

Currently, we are dealing with a key issue for our domain experts, namely the
management of changing names of, e.g., institutions that changed name retaining

---

[3] http://www.openlinksw.com/.

the same functions, across time and space. This particular point still represents an open challenge in this application domain – see [3].

Furthermore, we are developing a data integration layer in order to exploit information coming from relevant external sources, i.e. DBpedia [9] and the Ontology of the Chamber of Deputies, and integrate them in the STOLE ontology.

Concerning data navigation and visualization, we also intend to offer different ways to browse the STOLE resources, e.g., using interactive maps and LodLive[4] data graph representation.

We are also designing a Graphical User Interface to support the ontology population stage, in order to improve this process. More, concerning the ontology population, we are studying solutions for its automatization on the basis of some recent contributions – see, e.g., [10–12]. Finally, once the ontology will be fully populated, we are planning to perform an experimental analysis on the STOLE ontology involving state of the art DL reasoners on both classification and query answering tasks.

Acknowledgments. The authors wish to thank the anonymous reviewers for their valuable comments and suggestions to improve the paper. The authors would also like to thank Dott. Salvatore Mura and Prof. Francesco Soddu for the valuable discussions about the application domain. This work has been partially supported by MIUR.

References

1. Kruk, S.R., Westerki, A., Kruk, E.: Architecture of semantic digital libraries. In: McDaniel, B., Krik, S.R. (eds.) Semantic Digital Libraries, pp. 77–85. Springer, Berlin (2009)
2. Ahonen, E., Hyvonen, E.: Publishing historical texts on the semantic web-a case study. In: 2009 IEEE International Conference on Semantic Computing, ICSC 2009, pp. 167–173. IEEE (2009)
3. Meroño-Peñuela, A., Ashkpour, A., van Erp, M., Mandemakers, K., Breure, L., Scharnhorst, A., Schlobach, S., van Harmelen, F.: Semantic technologies for historical research: a survey. Semantic Web Journal [under review] (2012). http://www.semantic-web-journal.net/sites/default/files/swj301.pdf
4. Grau, B.C., Horrocks, I., Motik, B., Parsia, B., Patel-Schneider, P., Sattler, U.: Owl 2: the next step for owl. Web Seman. Sci. Serv. Agents World Wide Web **6**(4), 309–322 (2008)
5. Gennari, J.H., Musen, M.A., Fergerson, R.W., Grosso, W.E., Crubézy, M., Eriksson, H., Noy, N.F., Tu, S.W.: The evolution of protégé: an environment for knowledge-based systems development. Int. J. Hum. Comput. Stud. **58**(1), 89–123 (2003)
6. Horridge, M., Bechhofer, S.: The owl api: a java api for owl ontologies. Seman. Web **2**(1), 11–21 (2011)
7. Shearer, R., Motik, B., Horrocks, I.: Hermit: A highly-efficient owl reasoner. In: OWLED, vol. 432 (2008)

---

[4] http://lodlive.it.

8. Huynh, D.F., Karger, D.R., Miller, R.C.: Exhibit: lightweight structured data publishing. In: Proceedings of the 16th International Conference on World Wide Web, pp. 737–746. ACM (2007)
9. Bizer, C., Lehmann, J., Kobilarov, G., Auer, S., Becker, C., Cyganiak, R., Hellmann, S.: Dbpedia - a crystallization point for the web of data. Web Seman. Sci. Serv. Agents World Wide Web **7**(3), 154–165 (2009)
10. Fernández, M., Cantador, I., López, V., Vallet, D., Castells, P., Motta, E.: Semantically enhanced information retrieval: an ontology-based approach. Web Seman. Sci. Serv. Agents World Wide Web **9**(4), 434–452 (2011)
11. Kara, S., Alan, Ö., Sabuncu, O., Akpınar, S., Cicekli, N.K., Alpaslan, F.N.: An ontology-based retrieval system using semantic indexing. Inf. Syst. **37**(4), 294–305 (2012)
12. Sánchez, D., Batet, M., Isern, D., Valls, A.: Ontology-based semantic similarity: a new feature-based approach. Expert Syst. Appl. **39**(9), 7718–7728 (2012)

Semantic Views of Homogeneous Unstructured Data

Weronika T. Adrian[1,2], Nicola Leone[1], and Marco Manna[1($\boxtimes$)]

[1] Department of Mathematics and Computer Science,
University of Calabria, Cosenza, Italy
manna@mat.unical.it
[2] AGH University of Science and Technology,
Al.A.Mickiewicza 30, Krakow, Poland

Abstract. *Homogeneous unstructured data* (HUD) are collections of unstructured documents that share common properties, such as similar layout, common file format, or common domain of values. Building on such properties, it would be desirable to automatically process HUD to access the main information through a semantic layer – typically an ontology – called *semantic view*. Hence, we propose an ontology-based approach for extracting semantically rich information from HUD, by integrating and extending recent technologies and results from the fields of classical information extraction, table recognition, ontologies, text annotation, and logic programming. Moreover, we design and implement a system, named KnowRex, that has been successfully applied to curriculum vitae in the Europass style to offer a semantic view of them, and be able, for example, to select those which exhibit required skills.

Keywords: Unstructured data · Ontologies · Semantic information extraction · Table recognition · Semantic views

1 Introduction

Context and Motivation. By its nature, the Web has been conceived as an enormous distributed source of information which behaves as an open system to facilitate data sharing. However, the concrete way how the Web has been populated gave rise to a large amount of knowledge which is accessible only to humans but not to computers. A large slice of this knowledge is destined to remain only human-readable. But there is another relevant portion of it which could be automatically manipulated to be processed by computers. This is the case, for example, of *homogeneous unstructured data* (HUD) which are collections of unstructured documents that share common properties, such as similar layout, common file format, or common domain of values, just to mention a few.

Building on their common properties, it would be desirable to automatically process HUD to access the main information they contain through a semantic layer which is typically given in the form of an ontology, and that we call *semantic view*. The problem of identifying and extracting information from unstructured

© Springer International Publishing Switzerland 2015
B. ten Cate and A. Mileo (Eds.): RR 2015, LNCS 9209, pp. 19–29, 2015.
DOI: 10.1007/978-3-319-22002-4_3

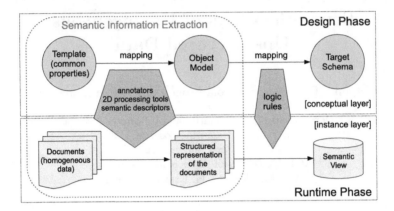

Fig. 1. Semantic Information Extraction with KnowRex.

documents is widely studied in the field of information and knowledge management and is referred to as Information Extraction (IE) [2,3,6]. However, most of the existing approaches to IE are mainly syntactic, and do not offer a uniform, clear, and semantic view of the relevant information.

Contribution. To offer a semantic view of a collection of HUD (even if encoded as pdf files), we propose and implement a system, named KnowRex, which splits the entire process in two different phases, called *design* and *runtime* (see Fig. 1). During the first one, the designer (*i*) defines the target schema for the semantic view of the original data, (*ii*) fixes an object model to offer a structured representation of the documents, (*iii*) arranges a suite of annotation units (such as named entity extractors, natural language processing tools, and annotation tools based on thesauri or regular expressions) to define the "leaves" of the object model, (*iv*) chooses and calibrates one of the software programs that partition unstructured documents into two-dimensional grids, and (*v*) provides formal rules to structure the documents and to construct the semantic view. During the runtime phase, the system processes the documents as prescribed in the design phase to instantiate the object model first, and then the target schema. In particular, the process that provides a structured representation of the documents can be thought as a kind of IE task which is heavily driven by domain knowledge and semantics, while the process that constructs the semantic view from the structured version of the documents takes care of reorganizing the extracted knowledge to facilitate data analysis. To sharpen our system, we considered curriculum vitae in the *Europass* style to offer a semantic view of them and be able, for example, to select those which exhibit required skills. The main contributions of the paper are:

▶ We present an ontology-based approach to IE which allows for extracting semantically rich information from unstructured data sharing some common features. To this end, we integrate and extend recent technologies and results from the fields of *classical information extraction*, *table recognition*, *ontologies*, *text annotation*, and *logic programming*.

▶ We design and implement a system, named KnowRex, which realizes our ontology-based approach to IE, and provides access to HUD via semantic views.

▶ On the application side, we have successfully applied KnowRex to offer a semantic view to curriculum vitae in the Europass style.

Related Work. The literature of the academic and commercial worlds offers a variety of approaches and tools to IE that are either programmed manually, or learned by semi-automatically supervised systems while a user interacts with example documents. A main shortcoming of these approaches, however, is their lack of understanding of extracted information. More recently, some works have shown the promise of deducing and encoding formal knowledge in the form of ontologies [1,5,7,8]. These approaches use ontologies either to improve the extraction phase as a way to present the results of the extraction, or to allow matching different representations across sources. The notion of semantic descriptors introduced in Sect. 3 has been inherited from HiLeX [9]. However, we have refined and extended their shape. Our approach follows the line of combining different techniques [4,10] to obtain comprehensive results. In KnowRex we include the following annotation tools: Alchemy (http://www.alchemyapi.com), DBpedia Spotlight (http://dbpedia.org/spotlight), Extractiv (http://extractiv.com), OpenCalais (http://www.opencalais.com), Lupedia (http://www.old.ontotext.com/lupedia), and StanfordNER (http://nlp.stanford.edu/software/CRF-NER.shtml).

2   Automatic Curriculum Analysis

For testing our framework, we selected the European standard style for Curriculum Vitae documents called Europass (see Fig. 2). This choice ensures that the input documents have similar two-column layout and organization of data. Despite some differences between single Europass CV documents, they can be seen as a collection of HUD, and we can assume a certain *template*, which consists in: (*i*) two-column layout, (*ii*) same file format (actually pdf, containing no information about sections/subsections which must be reconstructed at runtime), (*iii*) fixed set of labels (in the left column), (*iv*) common domain of values (personal information, education, work experience etc.).

For the target database, we are allowed to describe only the portion of information considered relevant. Let us assume the following target schema for the considered use case: candidate(Id, Name, Surname, Phone, Email, Address, Gender, Nationality, License); workExperience(Id, Company, BusinessSector, StartDate, EndDate); candWE(IdCandidate, IdWorkExperience).

The problem of recruiters and the goal of our system is to extract appropriate information from a collection of documents, and enter it into the target database. Some information can be localized by identifying appropriate sections and labels in the left column (e.g., name, surname, address etc.). Also, a part from the driving license, all the information needed for relation candidate are grouped together. For other information, it may be necessary to combine the knowledge about the structure of the input with the semantics of data. For instance, for

Personal information	
First name(s) / Surname(s)	**Weronika Adrian**
Address(es)	Via Pietro Bucci
Telephone(s)	+39 123 456 7890
E-mail	
Nationality	Polish
Gender	Female

Work experience	
Dates	10.2009-present
Occupation or position held	Research & Teaching Assistant
Main activities and responsibilities	Research: Semantic Technologies, Logic Programming, Knowledge Engineering; Teaching: Unix/GNU/Linux, Algorithms and Data Structures, Operating Systems, Semantic Web
Name and address of employer	AGH University of Science and Technology
Type of business or sector	Science and education
Dates	8.2008-9.2008
Occupation or position held	Software Developer
Main activities and responsibilities	Web Programming (html, css, javascript, php, ajax, mysql)
Name and address of employer	HolidayCheck AG
Type of business or sector	Information Technology

Fig. 2. Fragment of a Curriculum Vitae PDF-document in the Europass style.

work and education, it is possible to locate the institutions by analyzing the labels in the first column and extracting everything that follows them on the right. However, it would be beneficial to use also the *semantic annotators* that can recognize particular phrases as names of schools or companies to get more precise information. Finally, some information may be dispersed through the document. For instance, one may want to extract information about the candidates' skills, and this may be given in several ways. There exist dedicated sections in the Europass style, but these are not always filled by candidates. Thus, the skills may be also extracted from the other parts of the document. For instance, one may recognize "practical skills" as gained during the work experience, and "theoretical skills", if they are listed within the education section.

3   The Design Phase in KnowRex

During the design phase, a project is configured to perform operations on a collection of HUD, to obtain information desired by user, in a particular form. To do it, the designer should first identify the template (see Sect. 2), and then construct the target schema. Based on the template, they define an *object model* of the input, concepts of which will be recognized by different tools: two-dimensional processing tools, annotators, and semantic descriptors. Then, they configure the system and arranges the external tools so that the object model can be built. Afterwards, the designer writes logic rules that map the object model into target schema. The result of the design phase is used at runtime to process the actual documents to create the semantic view.

Definition of the Target Schema. This step is crucial for the definition of a desired output of the system. In fact, the designer has to decide how to organize the information that will be extracted from the documents. Two main options are allows: target schema may be either a relational database or an ontology. And this schema should be consistent and realistic i.e., it should be easy to populate it manually, only by analyzing the input documents. In our running example, the definition of the target schema has been given in Sect. 2.

Definition of the Object Model. By considering the template and the target schema, the designer should fix an object model for some HUD, which consists of a hierarchical forest-like structure. To define it, we uses the ontology language OntoDLP [11] that offers a good balance between ontological and logic-programming features. In OntDLP, one can define both object types, and relation types to express relationships between objects. Object types are preceded by keyword **entity**, and the subclass relationship is expressed via the term **isa**. Objects may have zero or more attributes which are specified in the type definition, by giving their names and types. By default, a class inherits attributes from its superclass. Relation types can be defined by keyword **relation**, and giving a name and attributes for this relation (see Sect. 4 for examples). One can also state assertions about objects.

Within the object model, a few types of objects may be identified. First, there are concepts that belong to an ontology representation of a document. This representation is independent of the use case, it is present in KnowRex by default and does not need any configuration. It contains one-dimensional objects such as token (basic elements of text) and delimiters, such as start and end of a line. It also provides two-dimensional concepts, empty and filled cells, to represent the basic elements of the document structure, e.g.: **entity** ontologyObject. **entity** 1DObject **isa** ontologyObject. **entity** token **isa** 1DObject. **entity** delimiter **isa** 1DObject. **entity** 2DObject **isa** ontologyObject. **entity** cell **isa** 2DObject. **entity** emptyCell **isa** cell. **entity** filledCell **isa** cell(value:**string**).

The second group of concepts is constituted by the categories that can be identified within the content of the document. For a CV use case, we can think of places, persons, companies, schools, skills, different professional terms, e.g. names of programming languages, languages etc. This set of concepts is defined by a designer and heavily depends on the use case, e.g.: **entity** semanticCategory(value:**string**). **entity** person **isa** semanticCategory. **entity** place **isa** semanticCategory. **entity** educationInstitution **isa** semanticCategory.

Finally, there is a group of concepts that describe domain-dependent elements of the structure of the document. These are concepts typically appearing in the considered HUD, such as section headlines, typical labels etc. These concepts are also given by a designer, e.g.: **entity** domainObject. **entity** eucv_label **isa** domainObject. **entity** eucv_name_label **isa** eucv_label. **entity** eucv_label_box **isa** domainObject. **entity** eucv_name_label_box **isa** eucv_label_box.

Arrangement of the Semantic Annotators. In this step, the designer selects the annotators to be used, then chooses classes that should be searched for, and configures each annotator: provides a mapping from the tool's output to the object model, and sets the tool's specific properties. In the case of Europass CV analysis, we have selected: OpenCalais, StanfordNER, Lupedia, a custom annotator for recognizing e-mail addresses and dates, and a label annotator based on pattern recognition that recognizes labels typical for Europass CV. Decisions about the arrangement of annotators are made by tries and errors on exemplary input data. Sometimes, it is beneficial to use more than one tool for recognizing the same category. The resulting potential redundancy is not harmful, instead the recall of extraction may improve.

Two-dimensional Document Analysis. Knowing the context in which certain phrase appears is helpful for semantic information extraction. In some input data formats, e.g. pdf documents, the information about the structure is lost; while visible to human eye, it is not obvious for a machine. Thus, we need to recover the structure to obtain a meaningful representation of the input documents. To this end, this step configures an external *two-dimensional processor* and a *refinement module* inside KnowRex. As a two-dimensional processor, we have used Quablo (http://www.quablo.eu/) that can recognize a set of regular tables within a pdf document. The representation obtained from this tool is then improved by a special module that works with domain concepts, such as labels of the Europass template. The module produces improved structure, merging appropriate cells (for example, if a label spans across two cells, these cells will be merged). Finally, one- and two-dimensional tokenizers (tools inside the KnowRex core) are used to identify the basic one- and two-dimensional objects of the document. In the end, we obtain a grid representation of the document that consists of two-dimensional objects (cells) containing one-dimensional one (text fragments, delimiters).

Semantic Descriptors Specification. While the semantic annotators identify single words or phrases as belonging to specific classes (producing the "leaves" of the object model), and the two-dimensional processing adds structure to the input, the semantic descriptors can combine and use the above information to build more complex objects. Semantic descriptors are rules that organize two-dimensional and one-dimensional objects into *descriptions* to extract additional information. This is done on several levels. To help the intuition, we illustrate the semantic descriptors by examples.

First, a designer should identify parts of the document that will help to localize other data portions, e.g.:

```
<eucv_email_label_box()> ::- <filledCell()> CONTAINS <eucv_email_label()>
```

With this simple descriptor, we intend to create a (two-dimensional) concept eucv_email_label_box that defines a cell in which there is a (one-dimensional) eucv_email_label. The object we want to extract always resides on the left-hand-side of the operator "::-" (in the head of a descriptor), while on the right (in the body), there are objects that must be found in order to create it. In this example, we look for a cell within which there is a particular domain concept, an `eucv_email_label`. If we find a cell with this label inside, the cell can be recognized as a `eucv_email_label_box`.

Descriptors can join several cells that appear in a document one after another (horizontally or vertically). This is useful, if we want to say that there exist a particular object, if there is a specific sequence of cells:

```
<candidateEmail(E)> ::- <eucv_email_label_box()>
                        (<filledCell(X)> CONTAINS <email(X)> {E:=X;})
```

This description should be read as: "A `candidateEmail` is a two-dimensional object that captures two cells: the first is an `eucv_email_label_box` and is followed (horizontally) by a `filledCell` that contains a (one-dimensional) object `email` with value X. The new object spans across both cells, and the value of the object `email` becomes the value of `candidateEmail`." By using the context (first there is a box with an e-mail label, and then there is a cell with an e-mail address), we ensure that, even if the CV contains a few e-mail addresses, we select the correct one, because if the e-mail address appears in this place, it must be in the Personal Information section and thus, it is the e-mail of the candidate.

We can also aggregate the concepts and attributes extracted by other semantic descriptors to build more complex ones:

```
<personalInformation(N, S, A, P, E, Nt G)> ::|
                <candidateName(X)> {N:=X;} <candidateSurname(X)> {S:=X;}
                <candidateAddress(X)> {A:=X;} <candidatePhone(X)> {P:=X;}
        <candidateNationality(X)> {Nt:=X;} <candidateGender(X)> {G:=X;}
```

This semantic descriptor aggregates results of other descriptors that extract single information about a candidate. It describes a sequence of concepts that must appear one after another vertically, which we mark with the "::|" operator. This aggregation must reflect the order in which information is given in the documents. The line breaks within a descriptor do not influence its semantics.

Within cells, we can create complex one-dimensional objects by using one-dimensional operator "::", a recurrence structure "(*sequence of terms*)+" and a keyword "..." that allows to skip some objects, e.g.:

```
<list_of_skills(S)> :: {S:=[];} <startOfLine> ...
                    (<IndustryTerm(S1)> {S&=S1;} ...)+ <endOfLine>
```

This descriptor works for one-dimensional objects that are all located in one cell (treated as a single line thanks to the two-dimensional processing). Here, we want to create a list, so we initialize the attribute `S:=[]`. Then, we look for a concept `IndustryTerm`, recognized by a semantic annotator, append its attribute value to S({S&=S1;}), and place the term in a recurrence structure. The expression (`<IndustryTerm(S1)>`{S&=S1;} ...)+ means that there may be some objects after the `IndustryTerm` that we ignore, and if we find another object `IndustryTerm`, we append its attribute value to the list again. By using the keyword "..." before the recurrence, we say that we can skip some objects, i.e., the recurrence structure may appear anywhere between the `startOfLine` and the `endOfLine`. The descriptor creates a new object `list_of_skills` that stores as an attribute a list of `IndustryTerm` objects' attributes.

Finally, semantic descriptors may use the information about the placement of objects within the document (e.g. presence of a given object within specific section) to produce new objects that are not explicitly present in text, e.g.:

```
<list_of_practical_skills(S)> ::- <eucv_work_act_resp_label_box()>
                (<filledCell(X)> CONTAINS <list_of_skills(X)> {S:=X;})
```

In this example, we use a `eucv_work_act_resp_label_box`, a domain concept that represents a cell containing "Activities and Responsibilities" label for selected Work Experience. This way, we look for lists of skills present only within the Work Experience subsections (and not for example within Education ones) and we can call them practical skills.

From Object Model to Target Schema. The design phase in KnowRex is completed with the definition of a mapping from object model classes to the concepts of the target schema. This mapping, written in a form of Datalog rules, is used to automatically create a semantic view of the (structured) input documents during the runtime phase. In the head of rules, there are concepts from the target schema, and in the body – objects from the object model (and auxiliary objects such as candidate ID). Partial mapping of the object model presented in Sect. 3 to the target schema shown in Sect. 2 is as follows:

```
candidate(Id,N,S,P,E,A,G,Nt,D) :- ID:cv_candidate_id(Id),
        PI:personalInformation(N,S,A,P,E,Nt,D,G),
        CDL:candidateDrivingLicence(D).
workExperience(Id, Company, BusinessSector, Start, End) :-
        WEID:work_experience_id(Id), C:company(Company,BusinessSector),
        WED:workExperienceDates(Start,End).
```

During runtime, the rules instantiated for actual objects extracted from the documents produce instances of the target schema. When the design phase ends, all the configuration information is merged into a project main configuration file.

4   The Runtime Phase of the System

Once the design of the project is done, KnowRex can be run over a collection of HUD. In the first stage of the document analysis, the two-dimensional structure of the document is recognized. First, a two-dimensional processor is used. Its output is then refined according to domain knowledge (specific labels, structure elements or keywords). Subsequently, this improved structure is analyzed by one- and two-dimensional tokenizers, tools hidden from a user, that identify the atomic one- and two-dimensional components of a document (tokens and cells).

KnowRex uses internal two-dimensional representation of objects that helps localize them within the documents. For each two-dimensional object, a relation biPosition is added that specify the row and column on the document "grid", where the object appears. For all one-dimensional objects (that are located inside the two-dimensional cells), two relations are added: belongsTo that identifies the containing cell by its id, and onePosition which denotes the position of the object within a cell. The definitions of these relations in OntoDLP are as follows: **relation** position(obj:ontologyObject, start:**int**, end:**int**). **relation** onePosition(obj:1DObject, start:**int**, end:**int**). **relation** biPosition(obj:2DObject, xs:**int**, ys:**int**, xe:**int**, ye:**int**). **relation** belongsTo(obj:1DObject, obj2:2DObject).

At the end of the two-dimensional processing stage, an ontological model of the document is obtained. It contains information about positions of the one- and two-dimensional objects within the document. The whole document is divided into cells, and the biPosition relations denote the coordinates (row and column) of the beginning and end of each cell, e.g.: filled19:filledCell('anna@w3.org'). biPosition(filled19, 1, 8, 2, 9). For one-dimensional objects, we have the following instances: tk123:token('manager'). onePosition(tk123, 0, 6). belongsTo(tk123, filled80). tk124:token('of'). onePosition(tk124, 6, 7). belongsTo(tk124, filled80). This representation is *normalized* i.e., the positions of blank spaces are omitted and the tokens follow one another. Such a representation is a reference for semantic annotators that may treat blank spaces differently.

Next, there is the annotation stage, in which selected semantic annotators are run over the identified cells and label the parts of text as objects belonging to different classes (such as Places, Persons, IndustryTerms, etc.) The representation of the identified objects (new logic facts that carry information about the annotator that found the object) is added to the fact base, e.g.: ann2:email('anna@w3.org'). onePosition(ann2, 0, 10). belongsTo(ann2, filled19).

Once the annotation stage is finished, the semantic descriptors which have been compiled into logic rules are executed over the facts representing the objects within a document. Each descriptor is transformed into a set of logical rules that first extract the portion of the document complying to the descriptor body, and then create a new object, specified in the descriptor head. At the end, a new object in OntoDLP is created, together with its one- or two-dimensional position (and optionally, *belonging to* a cell, if it is a one-dimensional object).

Finally, the identified objects of the object model are transformed into the instances of the semantic view. With use of the mapping defined in the design phase, the target schema is populated with instances extracted from the input

Candidate table:

Id	Name	Surname	Phone	Email	Address	Gender	Nationality	Driving License
12	Weronika	Adrian	+39 123 456 7890	w.adrian@mat.unical.it	Via Pietro Bucci	F	PL	B
13	Anna	Falcone	+48 987 654 321	anna@w3.org	-	F	IT	A

WorkExperience table:

Id	Company	Business Sector	Start Date	End Date
1	The International School of Kraków	Education	2006	2008
2	HolidayCheck AG	Information technology	2008	2008
3	AGH University of Science and Technology	Science and education	2009	-
4	World Wide Web Consortium (W3C)	Information technology	2010	2012

CandWE table:

Id Candidate	Id Work Experience
12	1
12	2
12	3
13	4

Fig. 3. A fragment of the semantic view of an Europass curriculum vitae.

document. Technically, this is done by additional logic rules that transform the objects to the target representation. Fragment of the semantic view for a set of two test CV documents is given in Fig. 3.

5 Discussion and Conclusion

We have described an ontology-based approach for extracting and organizing semantically rich information from HUD. This approach has been implemented in a system called KnowRex, which has been tested on curricula in the Europass style, stored as pdf files. Roughly, the design phase has been carried out in two man-weeks. From our preliminary analysis over 80 CVs, it appeared that the two-dimensional structure recognition and the recall of third-party annotators are the main bottlenecks. With initial configuration for the Europass template, Quablo worked well for about 50 % of documents. For further 20 % of documents, satisfying results were obtained by small adjustments of the tool (margin toleration etc.). While the precision of the semantic annotators is satisfying, their sometimes low recall may be compensated by adjusting home-made dictionary-based annotators. Logical rules (semantic descriptors and mapping rules) worked as expected on the found objects without loss of precision. KnowRex is sufficiently flexible and modular to be suitable for various scenarios in which HUD are available. We are currently applying the system on Wiki sites to automatically generate Semantic Wiki versions of them.

Acknowledgements. The work has been supported by Regione Calabria, programme POR Calabria FESR 2007–2013, within project "KnowRex: Un sistema per il riconoscimento e l'estrazione di conoscenza".

References

1. Anantharangachar, R., Ramani, S., Rajagopalan, S.: Ontology guided information extraction from unstructured text. CoRR abs/1302.1335 (2013)
2. Balke, W.T.: Introduction to information extraction: basic notions and current trends. Datenbank-Spektrum **12**(2), 81–88 (2012)
3. Chang, C.H., Kayed, M., Girgis, M.R., Shaalan, K.F.: A survey of web information extraction systems. IEEE Trans. Kn. Data Eng. **18**(10), 1411–1428 (2006)
4. Chen, L., Ortona, S., Orsi, G., Benedikt, M.: Aggregating semantic annotators. In: Proceedings VLDB Endow, vol. 6 no. 13, pp. 1486–1497 (2013)
5. Furche, Tim, Gottlob, Georg, Grasso, Giovanni, Orsi, Giorgio, Schallhart, Christian, Wang, Cheng: Little knowledge rules the web: domain-centric result page extraction. In: Rudolph, Sebastian, Gutierrez, Claudio (eds.) RR 2011. LNCS, vol. 6902, pp. 61–76. Springer, Heidelberg (2011)
6. Jiang, J.: Information extraction from text. In: Aggarwal, C.C., Zhai, C.X. (eds.) Mining Text Data, pp. 11–41. Springer, US (2012)
7. Kara, S., Alan, O., Sabuncu, O., Akpinar, S., Cicekli, N.K., Alpaslan, F.N.: An ontology-based retrieval system using semantic indexing. Inf. Syst. **37**(4), 294–305 (2012)

8. Karkaletsis, Vangelis, Fragkou, Pavlina, Petasis, Georgios, Iosif, Elias: Ontology based information extraction from text. In: Paliouras, Georgios, Spyropoulos, Constantine D., Tsatsaronis, George (eds.) Multimedia Information Extraction. LNCS, vol. 6050, pp. 89–109. Springer, Heidelberg (2011)

9. Manna, M., Oro, E., Ruffolo, M., Alviano, M., Leone, N.: The H$\imath$LεX system for semantic information extraction. Trans. Large-Scale Data- Knowl.-Centered Syst. V **7100**, 91–125 (2012)

10. Mo, Qian, Chen, Yi-hong: Ontology-Based Web Information Extraction. In: Zhao, Maotai, Sha, Junpin (eds.) ICCIP 2012, Part I. CCIS, vol. 288, pp. 118–126. Springer, Heidelberg (2012)

11. Ricca, F., Leone, N.: Disjunctive logic programming with types and objects: The DLV$^+$ system. J. Appl. Logic **5**(3), 545–573 (2007)

Supportedly Stable Answer Sets for Logic Programs with Generalized Atoms

Mario Alviano[1](✉) and Wolfgang Faber[2]

[1] University of Calabria, Rende, Italy
alviano@mat.unical.it
[2] University of Huddersfield, Huddersfield, UK
wf@wfaber.com

Abstract. Answer Set Programming (ASP) is logic programming under the stable model or answer set semantics. During the last decade, this paradigm has seen several extensions by generalizing the notion of atom used in these programs. Among these, there are dl-atoms, aggregate atoms, HEX atoms, generalized quantifiers, and abstract constraints. In this paper we refer to these constructs collectively as generalized atoms. The idea common to all of these constructs is that their satisfaction depends on the truth values of a set of (non-generalized) atoms, rather than the truth value of a single (non-generalized) atom. Motivated by several examples, we argue that for some of the more intricate generalized atoms, the previously suggested semantics provide unintuitive results and provide an alternative semantics, which we call supportedly stable or SFLP answer sets. We show that it is equivalent to the major previously proposed semantics for programs with convex generalized atoms, and that it in general admits more intended models than other semantics in the presence of non-convex generalized atoms. We show that the complexity of supportedly stable answer sets is on the second level of the polynomial hierarchy, similar to previous proposals and to answer sets of disjunctive logic programs.

1 Introduction

Answer Set Programming (ASP) is a widely used problem-solving framework based on logic programming under the stable model semantics. The basic language relies on Datalog with negation in rule bodies and possibly disjunction in rule heads. When actually using the language for representing practical knowledge, it became apparent that generalizations of the basic language are necessary for usability. Among the suggested extensions are aggregate atoms (similar to aggregations in database queries) [2–5] and atoms that rely on external truth

---

The main ideas of this paper were also presented in [1]. Mario Alviano was partly supported by MIUR within project "SI-LAB BA2KNOW – Business Analitycs to Know", by Regione Calabria, POR Calabria FESR 2007-2013, within project "ITravel PLUS" and project "KnowRex", by the National Group for Scientific Computation (GNCS-INDAM), and by Finanziamento Giovani Ricercatori UNICAL.

B. ten Cate and A. Mileo (Eds.): RR 2015, LNCS 9209, pp. 30–44, 2015.
DOI: 10.1007/978-3-319-22002-4_4

valuations [6–9]. These extensions are characterized by the fact that deciding the truth values of the new kinds of atoms depends on the truth values of a set of traditional atoms rather than a single traditional atom. We will refer to such atoms as *generalized atoms*, which cover also several other extensions such as abstract constraints, generalized quantifiers, and HEX atoms.

Concerning semantics for programs containing generalized atoms, there have been several different proposals. All of these appear to coincide for programs that do not contain generalized atoms in recursive definitions. The two main semantics that emerged as standards are the PSP semantics [10–12], and the FLP semantics [13,14] (the latter coinciding with Ferraris stable models [15] for the language considered in this paper). In a recent paper [16] the relationship between these two semantics was analyzed in detail; among other, more intricate results, it was shown that the semantics coincide up to convex generalized atoms. It was already established earlier that each PSP answer set is also an FLP answer set, but not vice versa. So for programs containing non-convex generalized atoms, some FLP answer sets are not PSP answer sets. In particular, there are programs that have FLP answer sets but no PSP answer sets.

In this paper, we argue that the FLP semantics is still too restrictive, and some programs that do not have any FLP answer set should instead have answer sets. In order to illustrate the point, consider a coordination game that is remotely inspired by the prisoners' dilemma. There are two players, each of which has the option to confess or defect. Let us also assume that both players have a fixed strategy already, which however still depends on the choice of the other player as well. In particular, each player will confess exactly if both players choose the same option, that is, if both players confess or both defect. This situation can be represented using two propositional atoms for "the first player confesses" and "the second player confesses," which must be derived true when "both players choose the same option," a composed proposition encoded by a generalized atom. As will be explained later, the FLP semantics does not assign any answer set to a program encoding this scenario, and therefore also the PSP semantics will not assign any answer sets to such a program. We observe that such a program is also incoherent according to a more recent refinement of the FLP semantics [17], called *well-justified FLP*.

We point out that this is peculiar, as the scenario in which both players confess seems like a reasonable one; indeed, even a simple inflationary operator would result in this solution: starting from the empty set, the generalized atom associated with "both players choose the same option" is true; therefore, the atoms associated with "the first player confesses" and "the second player confesses" are derived true on the first application of the operator, which is also its fixpoint.

Looking at the reason why this is not an FLP answer set, we observe that it has two countermodels that prevent it from being an answer set, one in which only the first player confesses, and another one in which only the second player confesses (see Fig. 1). Both of these countermodels are models in the classical sense, but they are weak in the sense that they are not supported, meaning that there is no rule justifying their truth. This is a situation that does not

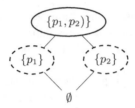

Fig. 1. Interpretations, supported (solid) and unsupported models (dashed) of the prisoners' dilemma example, where p_1 and p_2 are the propositions "the first player confesses" and "the first player confesses," respectively.

occur for programs without generalized atoms, which always have supported countermodels. We argue that one needs to look at supported countermodels, instead of looking at minimal countermodels. It turns out that doing this yields the same results not only for programs without generalized atoms, but also for programs containing convex generalized atoms, which we believe is the reason why this issue has not been noticed earlier.

This paper is first of all a position paper, in which we argue that the existing FLP and PSP semantics are too restrictive on the one hand, and that instead of defining restricting conditions, some conditions need be relaxed. We then proceed to define a new semantics along these lines and call it supportedly stable or SFLP (supportedly FLP) semantics. It provides answer sets for more programs than FLP and PSP, but is shown to be equal on convex programs. Analyzing the computational complexity of the new semantics, we show that it is in the same classes as the FLP and PSP semantics when considering polynomial-time computable generalized atoms. However, it should also be mentioned that the new semantics has its own peculiarities, for instance adding "tautological" rules like $p \leftarrow p$ can change the semantics of the program. These peculiarities suggest that a stronger notion of support is required for obtaining a solid semantics extending FLP.

The remainder of this paper is structured as follows. In Sect. 2, we present the notation and FLP semantics for programs with generalized atoms. After that, in Sect. 3 we analyze issues with the FLP semantics and define the SFLP semantics. In Sect. 4, we prove several useful properties of the new semantics. Finally, in Sect. 6, we discuss our results and provide outlines for future work.

2   Background

In this section we present the notation used in this paper and present the FLP semantics [13,14]. To ease the presentation, we will directly describe a propositional language here. This can be easily extended to the more usual ASP notations of programs involving variables, which stand for their ground versions (that are equivalent to a propositional program).

2.1 Notation

Let $\mathcal{B}$ be a countable set of *propositional atoms*. A *generalized atom* A on $\mathcal{B}$ is a pair (D_A, f_A), where $D_A \subseteq \mathcal{B}$ is the *domain* of A, and f_A is a mapping from 2^{D_A} to Boolean truth values $\{\mathbf{T}, \mathbf{F}\}$. To ease the presentation, we assume that the domain of each generalized atom is a finite set.

Example 1. Let p_1 represent the proposition "the first player confesses," and p_2 represent the proposition "the second player confesses." A generalized atom A representing the composed proposition "both players choose the same option" is such that $D_A = \{p_1, p_2\}$, $f_A(\{\}) = f_A(\{p_1, p_2\}) = \mathbf{T}$, and $f_A(\{p_1\}) = f_A(\{p_2\}) = \mathbf{F}$. ∎

A general rule r is of the following form:

$$H(r) \leftarrow B(r) \tag{1}$$

where $H(r)$ is a disjunction $a_1 \vee \cdots \vee a_n$ $(n \geq 0)$ of propositional atoms in $\mathcal{B}$ referred to as the head of r, and $B(r)$ is a generalized atom on $\mathcal{B}$ called the body of r. For convenience, $H(r)$ is sometimes considered a set of propositional atoms. A general program P is a set of general rules. Let $At(P)$ denote the set of propositional atoms occurring in P.

It should be noted that this is a very abstract notation, aiming to be general enough to encompass many concrete languages. Languages adopted in practical systems will feature concrete syntax in place of generalized atoms, for example aggregate atoms or dl-atoms. In the sequel, we will at times also use more concrete notation in examples to ease reading.

2.2 FLP Semantics

An *interpretation* I is a subset of $\mathcal{B}$. I is a *model* for a generalized atom A, denoted $I \models A$, if $f_A(I \cap D_A) = \mathbf{T}$. Otherwise, if $f_A(I \cap D_A) = \mathbf{F}$, I is not a model of A, denoted $I \not\models A$. I is a model of a rule r of the form (1), denoted $I \models r$, if $H(r) \cap I \neq \emptyset$ whenever $I \models B(r)$. I is a model of a program P, denoted $I \models P$, if $I \models r$ for every rule $r \in P$.

Note that the fact that rule bodies are forced to be a single generalized atom is not really a limitation, and will ease the presentation of the results in the paper. In fact, a single generalized atom is sufficient for modeling conjunctions, default negation, aggregates and similar constructs.

Example 2. A conjunction $p_1 \wedge \cdots \wedge p_n$ of $n \geq 1$ propositional atoms is equivalently represented by a generalized atom A such that $D_A = \{p_1, \ldots, p_n\}$, and $f_A(B) = \mathbf{T}$ if and only if $B = \{p_1, \ldots, p_n\}$.

A conjunction $p_1, \ldots, p_m, \sim p_{m+1}, \ldots, \sim p_n$ of literals, where $n \geq m \geq 0$, $p_1, \ldots, p_n$ are propositional atoms and $\sim$ denotes *negation as failure*, is equivalently represented by a generalized atom A such that $D_A = \{p_1, \ldots, p_n\}$, and $f_A(B) = \mathbf{T}$ if and only if $\{p_1, \ldots, p_m\} \subseteq B$ and $B \cap \{p_{m+1}, \ldots, p_n\} = \emptyset$.

An aggregate $COUNT(\{p_1, \ldots, p_n\}) \neq k$, where $n \geq k \geq 0$, and $p_1, \ldots, p_n$ are propositional atoms, is equivalently represented by a generalized atom A such that $D_A = \{p_1, \ldots, p_n\}$, and $f_A(B) = \mathbf{T}$ if and only if $|B \cap D_A| \neq k$.   ∎

In the following, when convenient, we will represent generalized atoms as conjunctions of literals or aggregate atoms. Subsets of $\mathcal{B}$ mapped to true by such generalized atoms will be those satisfying the associated conjunction.

Example 3. Consider the following rules:

$$r_1 : \; a \leftarrow COUNT(\{a, b\}) \neq 1 \qquad r_2 : \; b \leftarrow COUNT(\{a, b\}) \neq 1$$

The following are general programs that will be used for illustrating the differences between the semantics considered in this paper:

$$P_1 := \{r_1; r_2\} \qquad\qquad P_4 := \{r_1; r_2; a \vee b \leftarrow\}$$
$$P_2 := \{r_1; r_2; a \leftarrow b; b \leftarrow a\} \qquad P_5 := \{r_1; r_2; a \leftarrow\; \sim b\}$$
$$P_3 := \{r_1; r_2; \leftarrow\; \sim a; \leftarrow\; \sim b\}$$

Note that if a and b are replaced by p_1 and p_2, the aggregate $COUNT(\{a, b\}) \neq 1$ is equivalent to the generalized atom A from Example 1, and therefore program P_1 encodes the coordination game depicted in the introduction.   ∎

Generalized atoms can be partitioned into two classes, referred to as *convex* and *non-convex*, according to the following definition: A generalized atom A is convex if for all triples I, J, K of interpretations such that $I \subset J \subset K$, $I \models A$ and $K \models A$ implies $J \models A$. A convex program is a general program whose rules have convex bodies. Note that convex generalized atoms are closed under conjunction, but not under disjunction or complementation. In more detail, the conjunction of two generalized atoms A, A', denoted $A \wedge A'$, is such that $D_{A \wedge A'} = D_A \cup D_{A'}$, and for all $I \subseteq D_{A \wedge A'}$, $f_{A \wedge A'}(I) = f_A(I \cap D_A) \wedge f_{A'}(I \cap D_{A'})$. The disjunction $D_{A \vee A'}$ is defined similarly, and the complementation $\overline{A}$ of A is such that $D_{\overline{A}} = D_A$, and for all $I \subseteq D_A$, $f_{\overline{A}}(I) = \neg f_A(I)$. To show that convex generalized atoms are not closed under disjunction and complementation, an example is sufficient. Let A, A' be such that $D_A = D_{A'} = \{a, b\}$, $f_A(I) = \mathbf{T}$ if and only if $I = \emptyset$, and $f_{A'}(I) = \mathbf{T}$ if and only if $I = \{a, b\}$. Hence, A, A' are convex, but $A \vee A'$ is not. However, its complement $\overline{A \vee A'}$ is convex because true only for $\{a\}$ and $\{b\}$. Closure with respect to conjunction is proved by the following claim.

Lemma 1. *The conjunction $A \wedge A'$ of two convex generalized atoms is a convex generalized atom.*

Proof. Let $I \subset J \subset K$ be such that $I \models A \wedge A'$ and $K \models A \wedge A'$. Hence, $I \models A$, $K \models A$, $I \models A'$, and $K \models A'$ by definition of $A \wedge A'$. Since A and A' are convex, we have $J \models A$ and $J \models A'$, which in turn imply $J \models A \wedge A'$.   □

We now describe a reduct-based semantics, usually referred to as FLP, which has been introduced and analyzed in [13, 14].

Definition 1 (FLP Reduct). *The FLP reduct P^I of a program P with respect to I is defined as the set $\{r \in P \mid I \models B(r)\}$.*

Definition 2 (FLP Answer Sets). *I is an FLP answer set of P if $I \models P$ and for each $J \subset I$ it holds that $J \not\models P^I$. Let $FLP(P)$ denote the set of FLP answer sets of P.*

Example 4. Consider the programs from Example 3:

- The models of P_1 are $\{a\}$, $\{b\}$ and $\{a,b\}$, none of which is an FLP answer set. Indeed, $P_1^{\{a\}} = P_1^{\{b\}} = \emptyset$, which have the trivial model $\emptyset$, which is of course a subset of $\{a\}$ and $\{b\}$. On the other hand $P_1^{\{a,b\}} = P_1$, and so $\{a\} \models P_1^{\{a,b\}}$, where $\{a\} \subset \{a,b\}$. We will discuss in the next section why this is a questionable situation.
- Concerning P_2, it has one model, namely $\{a,b\}$, which is also its unique FLP answer set. Indeed, $P_2^{\{a,b\}} = P_2$, and hence the only model of $P_2^{\{a,b\}}$ is $\{a,b\}$.
- Interpretation $\{a,b\}$ is also the unique model of program P_3, which however has no FLP answer sets. Here, $P_3^{\{a,b\}} = P_1$, hence similar to P_1, $\{a\} \models P_3^{\{a,b\}}$ and $\{a\} \subset \{a,b\}$.
- P_4 instead has two FLP answer sets, namely $\{a\}$ and $\{b\}$, and a further model $\{a,b\}$. In this case, $P_4^{\{a\}} = \{a \vee b \leftarrow\}$, and no proper subset of $\{a\}$ satisfies it. Also $P_4^{\{b\}} = \{a \vee b \leftarrow\}$, and no proper subset of $\{b\}$ satisfies it. Instead, for $\{a,b\}$, we have $P_4^{\{a,b\}} = P_4$, and hence $\{a\} \models P_4^{\{a,b\}}$ and $\{a\} \subset \{a,b\}$.
- Finally, P_5 has tree models, $\{a\}$, $\{b\}$ and $\{a,b\}$, but only one answer set, namely $\{a\}$. In fact, $P_5^{\{a\}} = \{a \leftarrow\sim b\}$ and $\emptyset$ is not a model of the reduct. On the other hand, $\emptyset$ is a model of $P_5^{\{b\}} = \emptyset$, and $\{a\}$ is a model of $P_5^{\{a,b\}} = P_1$.

Models and FLP answer sets of these programs are summarized in Table 1. ∎

3 SFLP Semantics

As noted in the introduction, the fact that P_1 has no FLP answer sets is striking. If we first assume that both a and b are false (interpretation $\emptyset$), and then apply a generalization of the well-known one-step derivability operator, we obtain truth of both a and b (interpretation $\{a,b\}$). Applying this operator once more again yields the same interpretation, a fix-point. Interpretation $\{a,b\}$ is also a supported model, that is, for all true atoms there exists a rule in which this atom is the only true head atom, and in which the body is true.

It is instructive to examine why this seemingly robust model is not an FLP answer set. Its reduct is equal to the original program, $P_1^{\{a,b\}} = P_1$. There are therefore two models of P_1, $\{a\}$ and $\{b\}$, that are subsets of $\{a,b\}$ and therefore inhibit $\{a,b\}$ from being an FLP answer set. The problem is that, contrary to $\{a,b\}$, these two models are rather weak, in the sense that they are not supported. Indeed, when considering $\{a\}$, there is no rule in P_1 such that a is the only true

Table 1. (Supported) models and (S)FLP answer sets of programs in Example 3, where A is the generalized atom $COUNT(\{a,b\}) \neq 1$.

	Rules		Models	FLP	Supported Models	SFLP
P_1	$a \leftarrow A$	$b \leftarrow A$	$\{a\}, \{b\}, \{a,b\}$	—	$\{a,b\}$	$\{a,b\}$
P_2	$a \leftarrow A$	$b \leftarrow A$	$\{a,b\}$	$\{a,b\}$	$\{a,b\}$	$\{a,b\}$
	$a \leftarrow b$	$b \leftarrow a$				
P_3	$a \leftarrow A$	$b \leftarrow A$	$\{a,b\}$	—	$\{a,b\}$	$\{a,b\}$
	$\leftarrow \sim a$	$\leftarrow \sim b$				
P_4	$a \leftarrow A$	$b \leftarrow A$	$\{a\}, \{b\}, \{a,b\}$	$\{a\}, \{b\}$	$\{a\}, \{b\}, \{a,b\}$	$\{a\}, \{b\}$
	$a \vee b \leftarrow$					
P_5	$a \leftarrow A$	$b \leftarrow A$	$\{a\}, \{b\}, \{a,b\}$	$\{a\}$	$\{a\}, \{a,b\}$	$\{a\}, \{a,b\}$
	$a \leftarrow \sim b$					

atom in the rule head and the body is true in $\{a\}$: The only available rule with a in the head has a false body. The situation for $\{b\}$ is symmetric.

It is somewhat counter-intuitive that a model like $\{a,b\}$ should be inhibited by two weak models like $\{a\}$ and $\{b\}$. Indeed, this is a situation that normally does not occur in ASP. For programs that do not contain generalized atoms, whenever one finds a $J \subset I$ such that $J \models P^I$ there is for sure also a K such that $J \subseteq K \subset I$, $K \models P^I$ and K is supported. Indeed, we will show in Sect. 4 that this is the case also for programs containing only convex generalized atoms. Our feeling is that since such a situation does not happen for a very wide set of programs, it has been overlooked so far.

We will now attempt to repair this kind of anomaly by stipulating that one should only consider supported models for finding inhibitors of answer sets. In other words, one does not need to worry about unsupported models of the reduct, even if they are subsets of the candidate. Let us first define supported models explicitly.

Definition 3 (Supportedness). *A model I of a program P is supported if for each $a \in I$ there is a rule $r \in P$ such that $I \cap H(r) = \{a\}$ and $I \models B(r)$. In this case we will write $I \models_s P$.*

Example 5. Continuing Example 4, programs P_1, P_2, and P_3 have one supported model, namely $\{a,b\}$. The model $\{a\}$ of P_1 is not supported because the body of the the rule with a in the head has a false body with respect to $\{a\}$. For a symmetric argument, model $\{b\}$ of P_1 is not supported either. The supported models of P_4, instead, are $\{a\}$, $\{b\}$, and $\{a,b\}$, so all models of the program are supported. Note that both models $\{a\}$ and $\{b\}$ have the disjunctive rule as the only supporting rule for the respective single true atom, while for $\{a,b\}$, the two rules with generalized atoms serve as supporting rules for a and b. Finally, the supported models of P_5 are $\{a\}$ and $\{a,b\}$. Supported models of these programs are summarized in Table 1. ∎

We are now ready to formally introduce the new semantics. In this paper we will normally refer to it as SFLP answer sets or SFLP semantics, but also call it *supportedly stable models* occasionally.

Definition 4 (SFLP Answer Sets). *I is an SFLP answer set of P if $I \models_s P$ and for each $J \subset I$ it holds that $J \not\models_s P^I$. Let $SFLP(P)$ denote the set of SFLP answer sets of P.*

Example 6. Consider again the programs from Example 3.

- Recall that P_1 has only one supported model, namely $\{a, b\}$, and $P_1^{\{a,b\}} = P_1$, but $\emptyset \not\models_s P_1^{\{a,b\}}$, $\{a\} \not\models_s P_1^{\{a,b\}}$, and $\{b\} \not\models_s P_1^{\{a,b\}}$, therefore no proper subset of $\{a, b\}$ is a supported model. Hence, it is an SFLP answer set.
- Concerning P_2, it has one model, namely $\{a, b\}$, which is supported and also its unique SFLP answer set. Indeed, recall that $P_2^{\{a,b\}} = P_2$, and hence no proper subset of $\{a, b\}$ can be a model (let alone a supported model) of $P_2^{\{a,b\}}$.
- Interpretation $\{a, b\}$ is the unique model of program P_3, which is supported and also its SFLP answer set. In fact, $P_3^{\{a,b\}} = P_1$.
- P_4 has two SFLP answer sets, namely $\{a\}$ and $\{b\}$. In this case, recall $P_4^{\{a\}} = \{a \vee b \leftarrow\}$, and no proper subset of $\{a\}$ satisfies it. Also $P_4^{\{b\}} = \{a \vee b \leftarrow\}$, and no proper subset of $\{b\}$ satisfies it. Instead, for $\{a, b\}$, we have $P_4^{\{a,b\}} = P_4$, hence since $\{a\} \models_s P_4^{\{a,b\}}$, and $\{b\} \models_s P_4^{\{a,b\}}$, we obtain that $\{a, b\}$ is not an SFLP answer set.
- Finally, P_5 has two SFLP answer sets, namely $\{a\}$ and $\{a, b\}$. In fact, $P_5^{\{a\}} = \{a \leftarrow \sim b\}$ and $P_5^{\{a,b\}} = P_1$.

The programs, models, FLP answer sets, supported models, and SFLP answer sets are summarized in Table 1. ∎

An alternative, useful characterization of SFLP answer sets can be given in terms of Clark's completion [18]. In fact, it is well-known that supported models of a program are precisely the models of its completion. We define this notion in a somewhat non-standard way, making use of the concept of generalized atom. We first define the completion of a propositional atom p with respect to a general program P as a generalized atom encoding the supportedness condition for p.

Definition 5. *The completion of a propositional atom $p \in \mathcal{B}$ with respect to a general program P is a generalized atom A such that $D_A = At(P)$, and for all $I \subseteq D_A$, $f_A(I) = \mathbf{T}$ if and only if $p \in I$ and there is no rule $r \in P$ for which $I \models B(r)$ and $I \cap H(r) = \{a\}$. Let $comp(p, P)$ denote the completion of p with respect to P.*

These generalized atoms are then used to effectively define a program whose models are the supported model of P.

Definition 6. *The completion of a general program P is a general program $comp(P)$ extending P with a rule $\leftarrow comp(p, P)$ for each propositional atom $p \in At(P)$.*

Example 7. Consider again the programs from Example 3.

– Program $comp(P_1)$ extends P_1 with the following rules:

$$\leftarrow a \wedge COUNT(\{a,b\}) = 1 \leftarrow b \wedge COUNT(\{a,b\}) = 1$$

– Program $comp(P_2)$ extends P_2 with the following rules:

$$\leftarrow a \wedge COUNT(\{a,b\}) = 1 \wedge \sim b \leftarrow b \wedge COUNT(\{a,b\}) = 1 \wedge \sim a$$

– Program $comp(P_3)$ is equal to $comp(P_1)$, and program $comp(P_4)$ extends P_4 with the following rules:

$$\leftarrow a \wedge COUNT(\{a,b\}) = 1 \wedge b \leftarrow b \wedge COUNT(\{a,b\}) = 1 \wedge a$$

– Program $comp(P_5)$ instead extends P_5 with the following rules:

$$\leftarrow a \wedge COUNT(\{a,b\}) = 1 \wedge b \leftarrow b \wedge COUNT(\{a,b\}) = 1$$

Note that the only model of $comp(P_1)$, $comp(P_2)$, and $comp(P_3)$ is $\{a,b\}$. As for $comp(P_4)$, and $comp(P_5)$, their models are $\{a\}$, $\{b\}$, and $\{a,b\}$. ∎

Proposition 1. *Let P be a general program, and I be an interpretation. Then, $I \models_s P$ if and only if $I \models comp(P)$.*

This characterization, which follows directly from [18], provides us with a means for implementation that relies only on model checks, rather than supportedness checks.

Proposition 2. *Let P be a general program, and I be an interpretation. Then, I is an SFLP answer set of P if $I \models comp(P)$ and for each $J \subset I$ it holds that $J \not\models comp(P^I)$.*

4   Properties

The new semantics has a number of interesting properties that we report in this section. First of all, it is an extension of the FLP semantics, in the sense that each FLP answer set is also an SFLP answer set.

Theorem 1. *Let P be a general program. Then, $FLP(P) \subseteq SFLP(P)$.*

Proof. Let I be an FLP answer set of P. Hence, each $J \subset I$ is such that $J \not\models P^I$. Thus, we can conclude that $J \not\models_s P^I$ for any $J \subset I$. Therefore, I is a SFLP answer set of P. □

The inclusion is strict in general. In fact, P_1 is a simple program for which the two semantics disagree (see Examples 3–6 and Table 1). On the other hand, the two semantics are equivalent for a large class of programs, as shown below.

Theorem 2. *If P is a convex program then $FLP(P) = SFLP(P)$.*

Proof. $FLP(P) \subseteq SFLP(P)$ holds by Theorem 1. For the other direction, consider an interpretation I not being an FLP answer set of P. Hence, there is $J \subset I$ such that $J \models P^I$. We also assume that J is a subset-minimal model of P^I, that is, there is no $K \subset J$ such that $K \models P^I$. We shall show that $J \models_s P^I$. To this end, suppose by contradiction that there is $p \in J$ such that for each $r \in P^I$ either $J \not\models B(r)$ or $J \cap H(r) \neq \{p\}$. Consider $J \setminus \{p\}$ and a rule $r \in P^I$ such that $J \setminus \{p\} \models B(r)$. Since $r \in P^I$, $I \models B(r)$, and thus $J \models B(r)$ because $B(r)$ is convex. Therefore, $J \cap H(r) \neq \{p\}$. Moreover, $J \cap H(r) \neq \emptyset$ because $J \models P^I$ by assumption. Hence, $(J \setminus \{p\}) \cap H(r) \neq \emptyset$, and therefore $J \setminus \{p\} \models P^I$. This contradicts the assumption that J is a subset-minimal model of P^I. $\square$

We will now focus on computational complexity. We consider here the problem of determining whether an SFLP answer set exists. We note that the only difference to the FLP semantics is in the stability check. For FLP, subsets need to be checked for being a model. For SFLP, instead, subsets need to be checked for being a supported model. Intuitively, one would not expect that this difference can account for a complexity jump, which is confirmed by the next result.

Theorem 3. *Let P be a general program whose generalized atoms are polynomial-time computable functions. Checking whether $SFLP(P) \neq \emptyset$ is in Σ_2^P in general; it is Σ_2^P-hard already in the disjunction-free case if at least one form of non-convex generalized atom is permitted. The problem is NP-complete if P is disjunction-free and convex.*

Proof. For the membership in Σ_2^P one can guess an interpretation I and check that there is no $J \subset I$ such that $J \models_s P$. The check can be performed by a coNP oracle. To prove Σ_2^P-hardness we note that extending a general program P by rules $p \leftarrow p$ for every $p \in At(P)$ is enough to guarantee that all models of any reduct of P are supported. We thus refer to the construction and proof by [16]. If P is disjunction-free and convex then $SFLP(P) = FLP(P)$ by Theorem 2. Hence, NP-completeness follows from results in [19]. $\square$

We would like to point out that the above proof also illustrates a peculiar feature of SFLP answer sets, which it shares with the supported model semantics: the semantics is sensitive to tautological rules like $p \leftarrow p$, as their addition can turn non-SFLP answer sets into SFLP answer sets.

5 Discussion

Existing semantics for logic programs with generalized atoms do not yield intended models under particular conditions. Specifically, the lack of intended models seems to be ascribable to unsupported countermodels. In the FLP semantics, the anomaly arises in connection with non-convex generalized atoms. As the example in the introduction shows, the anomaly affects other semantics as well. For example, it affects also description logic programs interpreted according to the recent well-justified FLP semantics [17], which selects among FLP answer sets. Former semantics for description logic programs, referred to as strong and weak stable models [7,8], behave differently. Without going into details, we use the prisoners' dilemma example from the introduction. Recall that there are two prisoners who will confess if (and only if) both players choose the same option. This situation can be represented using the following description logic program:

$$a(0) \leftarrow DL[A \barwedge a, B \barwedge b, A \uplus a, B \uplus b; Q](0)$$
$$b(0) \leftarrow DL[A \barwedge a, B \barwedge b, A \uplus a, B \uplus b; Q](0)$$

with the following ontology:

$$Q \equiv (A \sqcap B) \sqcup (\neg A \sqcap \neg B).$$

Here, $a(0)$ means that the first player confesses and $b(0)$ means that the second player confesses. The DL atoms transfer the extensions of a and b to A and B of the ontology, making sure that constants that are not in the extension of a will populate $\neg A$ (likewise for b and $\neg B$); this is achieved by means of the $\uplus$ and $\barwedge$ operators. The ontology defines Q, which represents "good" situations (both confess or both defect). The rule thus states that $a(0)$ and $b(0)$ should hold in a "good" situation. An alternative encoding using $\cupdot$ instead of $\uplus$ is given by the following program:

$$a(0) \leftarrow DL[A \barwedge a, B \barwedge b, C \cupdot a, D \cupdot b; Q](0)$$
$$b(0) \leftarrow DL[A \barwedge a, B \barwedge b, C \cupdot a, D \cupdot b; Q](0)$$

with the following ontology:

$$Q \equiv (\neg A \sqcap \neg B) \sqcup (\neg C \sqcap \neg D).$$

where operator $\cupdot$ increases $\neg C$ and $\neg D$ with the extension of a and b, respectively. These programs correspond to P_1 in Example 3, which just uses different notation. Interpretation $\{a(0), b(0)\}$ is both the only strong and weak answer set of the above programs. However, these semantics suffer from other problems. For example, if a fact $a(0)$ is added to the above programs, $\{a(0), b(0)\}$ would still be both a strong and weak answer set, along with $\{a(0)\}$, which we consider to be quite unintuitive.

Another semantics that selects among FLP answer sets, and is therefore affected by the anomaly on unsupported countermodels, is PSP [10–12]. Analyzing Fig. 2, we can better understand a common issue for PSP and well-justified FLP: The adopted operator for fixpoint computation cannot "jump" over gaps in the lattice of interpretations when establishing the truth value of a generalized atom. Examining the figure, the fact that $\{a(0)\}$ and $\{b(0)\}$ both do not satisfy the generalized atom prevents these operators from deriving $\{a(0), b(0)\}$. Yet these two interpretations are not relevant for the example in the introduction.

We observe that other interesting semantics, such as the one by [20], are also affected by the anomaly on unsupported models. In particular, the semantics by [20] is presented for programs consisting of arbitrary set of propositional

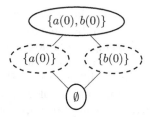

Fig. 2. Satisfying (solid) and unsatisfying (dashed) interpretations of the description logic programs in Sect. 5 encoding the prisoners' dilemma.

formulas, and it is based on a reduct in which false subformulas are replaced by $\perp$. Answer sets are then defined as interpretations being subset-minimal models of their reducts. For the notation used in this paper, when rewriting generalized atoms to an equivalent formula, the semantics by [20] coincides with FLP, which immediately shows the anomaly. In [20] there is also a method for rewriting aggregates, however $COUNT(\{a, b\}) \neq 1$ is not explicitly supported, but should be rewritten to $\neg(COUNT(\{a, b\}) = 1)$. Doing this, one can observe that for P_1, P_2, P_3, and P_5 the semantics of [20] behaves like SFLP (cf. Table 1), while for P_4 the semantics of [20] additionally has the answer set $\{a, b\}$, which is not a supported minimal model of the FLP reduct. P_4 therefore shows that the two semantics do not coincide, even if generalized atoms are interpreted as their negated complements, and the precise relationship is left for further study. However, we also believe that rewriting a generalized atom into its negated complement is not always natural, and we are also convinced that there should be a semantic difference between a generalized atom and its negated complement.

6 Conclusion

In this paper, we have first studied conditions under which existing semantics for logic programs with generalized atoms do not yield intended models, while they arguably should. Analyzing the reasons, we argue that this seems to be connected to a lack of support in countermodels. Motivated by this situation, we have then defined a new semantics for programs with generalized atoms, called supportedly stable models, supportedly FLP, or SFLP semantics. The new definition overcomes the anomaly on unsupported countermodels, and provides a robust semantics for programs with generalized atoms. We show several properties of this new semantics. For example, it coincides with the FLP semantics (and thus also the PSP semantics) on convex programs, and therefore also on standard programs. Furthermore, the complexity of common reasoning tasks is equal to the respective tasks using the FLP semantics. We also provide a characterization of the new semantics by a Clark-inspired completion.

Concerning future work, implementing a reasoner supporting the new semantics would be of interest, for example by compiling the new semantics in FLP, so to use current ASP solvers such as DLV [21], CMODELS [22], CLASP [23], and

WASP [24,25]. An application area would be systems that loosely couple OWL ontologies with rule bases, for instance by means of HEX programs. As we have shown earlier, HEX atoms interfacing to ontologies will in general not be convex, and therefore using them in recursive definitions falls into our framework, where the FLP and SFLP semantics differ. In fact, the solver dlvhex[1] does not produce any answer for the program and ontology provided in Sect. 5. We also believe that it would be important to collect example programs that contain non-convex generalized atoms in recursive definitions. We have experimented with a few simple domains stemming from game theory (as outlined in the introduction), but we are not aware of many other attempts. Our intuition is that such programs would be written in several domains that describe features with feedback loops, which applies to many so-called complex systems. Also computing or checking properties of neural networks might be a possible application in this area. Another area of future work arises from the fact that rules like $p \leftarrow p$ are not irrelevant for the SFLP semantics. To us, it is not completely clear how much of a drawback this really is. However, we intend to study this along two avenues. The first is trying to characterize programs for which this phenomenon occurs, the second is to propose variants of the SFLP semantics that do not exhibit it. Concerning the latter, one idea would be to require models or countermodels to have level mappings as defined in [17].

References

1. Alviano, M., Faber, W.: Semantics and compilation of answer set programming with generalized atoms. In: Konieczny, S., Tompits, H. (eds.) 15th International Workshop on Nonmonotonic Reasoning (NMR 2014), Number 1843-14-01 in INF-SYS Research Reports, Institut für Informationssysteme, pp. 214–222, July 2014
2. Niemelä, I., Simons, P., Soininen, T.: Stable model semantics of weight constraint rules. In: Gelfond, M., Leone, N., Pfeifer, G. (eds.) LPNMR 1999. LNCS, vol. 1730, pp. 317–331. Springer, Heidelberg (1999)
3. Niemelä, I., Simons, P.: Extending the smodels system with cardinality and weight constraints. In: Minker, J. (ed.) Logic-Based Artificial Intelligence, pp. 491–521. Kluwer Academic Publishers, Dordrecht (2000)
4. Dell'Armi, T., Faber, W., Ielpa, G., Leone, N., Pfeifer, G.: Aggregate functions in disjunctive logic programming: semantics, complexity, and implementation in DLV. In: Proceedings of the 18th International Joint Conference on Artificial Intelligence (IJCAI) 2003. Morgan Kaufmann Publishers, Acapulco, Mexico, pp. 847–852, August 2003
5. Faber, W., Pfeifer, G., Leone, N., Dell'Armi, T., Ielpa, G.: Design and implementation of aggregate functions in the DLV system. Theory Pract. Logic Program. **8**(5–6), 545–580 (2008)
6. Calimeri, F., Cozza, S., Ianni, G.: External sources of knowledge and value invention in logic programming. Ann. Math. Artif. Intell. **50**(3–4), 333–361 (2007)

---

[1] http://www.kr.tuwien.ac.at/research/systems/dlvhex/.

7. Eiter, T., Lukasiewicz, T., Schindlauer, R., Tompits, H.: Combining answer set programming with description logics for the semantic web. In: Principles of Knowledge Representation and Reasoning: Proceedings of the Ninth International Conference (KR2004), Whistler, Canada, pp. 141–151 (2004). Extended Report RR-1843-03-13, Institut für Informationssysteme, TU Wien, 2003

8. Eiter, T., Ianni, G., Lukasiewicz, T., Schindlauer, R., Tompits, H.: Combining answer set programming with description logics for the semantic web. Artif. Intell. **172**(12–13), 1495–1539 (2008)

9. Eiter, T., Ianni, G., Schindlauer, R., Tompits, H.: A uniform integration of higher-order reasoning and external evaluations in answer set programming. In: International Joint Conference on Artificial Intelligence (IJCAI) 2005, pp. 90–96, Edinburgh, UK, August 2005

10. Pelov, N.: Semantics of logic programs with aggregates. Ph.D. thesis, Katholieke Universiteit Leuven, Leuven, Belgium, April 2004

11. Pelov, N., Denecker, M., Bruynooghe, M.: Well-founded and Stable semantics of logic programs with aggregates. Theory Pract. Logic Program. **7**(3), 301–353 (2007)

12. Son, T.C., Pontelli, E.: A constructive semantic characterization of aggregates in ASP. Theory Pract. Logic Program. **7**, 355–375 (2007)

13. Faber, W., Leone, N., Pfeifer, G.: Recursive aggregates in disjunctive logic programs: semantics and complexity. In: Alferes, J.J., Leite, J. (eds.) JELIA 2004. LNCS (LNAI), vol. 3229, pp. 200–212. Springer, Heidelberg (2004)

14. Faber, W., Leone, N., Pfeifer, G.: Semantics and complexity of recursive aggregates in answer set programming. Artif. Intell. **175**(1), 278–298 (2011). Special Issue: John McCarthy's Legacy

15. Ferraris, P.: Logic programs with propositional connectives and aggregates. ACM Trans. Comput. Log. **12**(4), 25 (2011)

16. Alviano, M., Faber, W.: The complexity boundary of answer set programming with generalized atoms under the FLP semantics. In: Cabalar, P., Son, T.C. (eds.) LPNMR 2013. LNCS, vol. 8148, pp. 67–72. Springer, Heidelberg (2013)

17. Shen, Y., Wang, K., Eiter, T., Fink, M., Redl, C., Krennwallner, T., Deng, J.: FLP answer set semantics without circular justifications for general logic programs. Artif. Intell. **213**, 1–41 (2014)

18. Clark, K.L.: Negation as failure. In: Gallaire, H., Minker, J. (eds.) Logic and Data Bases, pp. 293–322. Plenum Press, New York (1978)

19. Liu, L., Truszczyński, M.: Properties and applications of programs with monotone and convex constraints. J. Art. Intell. Res. **27**, 299–334 (2006)

20. Ferraris, P.: Answer sets for propositional theories. In: Baral, C., Greco, G., Leone, N., Terracina, G. (eds.) LPNMR 2005. LNCS (LNAI), vol. 3662, pp. 119–131. Springer, Heidelberg (2005)

21. Alviano, M., Faber, W., Leone, N., Perri, S., Pfeifer, G., Terracina, G.: The disjunctive datalog system DLV. In: de Moor, O., Gottlob, G., Furche, T., Sellers, A. (eds.) Datalog 2010. LNCS, vol. 6702, pp. 282–301. Springer, Heidelberg (2011)

22. Lierler, Y., Maratea, M.: Cmodels-2: sat-based answer set solver enhanced to non-tight programs. In: Lifschitz, V., Niemelä, I. (eds.) LPNMR 2004. LNCS (LNAI), vol. 2923, pp. 346–350. Springer, Heidelberg (2003)

23. Gebser, M., Kaufmann, B., Schaub, T.: Conflict-driven answer set solving: From theory to practice. Artif. Intell. **187**, 52–89 (2012)

24. Alviano, M., Dodaro, C., Faber, W., Leone, N., Ricca, F.: WASP: a native ASP solver based on constraint learning. In: Cabalar, P., Son, T.C. (eds.) LPNMR 2013. LNCS, vol. 8148, pp. 54–66. Springer, Heidelberg (2013)
25. Alviano, M., Dodaro, C., Ricca, F.: Anytime computation of cautious consequences in answer set programming. TPLP **14**(4–5), 755–770 (2014)

Planning with Regression Analysis
in Transaction Logic

Reza Basseda$^{(\boxtimes)}$ and Michael Kifer

Stony Brook University, Stony Brook, NY 11794, USA
{rbasseda,kifer}@cs.stonybrook.edu

Abstract. Heuristic search is arguably the most successful paradigm in Automated Planning, which greatly improves the performance of planning strategies. However, adding heuristics usually leads to very complicated planning algorithms. In order to study different properties (e.g. completeness) of those complicated planning algorithms, it is important to use an appropriate formal language and framework. In this paper, we argue that Transaction Logic is just such a specification language, which lets one formally specify both the heuristics, the planning algorithm, and also facilitates the discovery of more general and more efficient algorithms. To illustrate, we take the well-known regression analysis mechanism and show that Transaction Logic lets one specify the concept of regression analysis easily and thus express *RSTRIPS*, a classical and very complicated planning algorithm based on regression analysis. Moreover, we show that extensions to that algorithm that allow indirect effects and action ramification are obtained almost for free. Finally, a compact and clear logical formulation of the algorithm lets us prove the completeness of *RSTRIPS*—a result that, to the best of our knowledge, has not been known before.

1   Introduction

Heuristic planning is an application of heuristic search to the domain of planning. To find a plan, a heuristic planner relies on the information about the regions of the search space in which a successful plan solution is likely to be found. Planning algorithms can use such information to guide their search and thus reduce the search space that will likely be explored in order to find a solution.

Most planners apply various sophisticated domain-independent heuristics [3,15,21,25]. Other approaches [1,11,22,26,29] use declarative formalisms (e.g. situation calculus or linear temporal logic) to express heuristic information. Declarative heuristic information can be used to prune the search space [1,26] or steer the search in promising directions [13,18]. Although the declarative representation of heuristic information provides multiple advantages for specifying and generalizing heuristics, many of the above approaches do not get as much attention as the heuristic methods. There are several reasons for this:

- Declarative approaches typically propose a framework where a user can specify some heuristics but they are not flexible enough for representing

This work was supported, in part, by the NSF grant 0964196.

B. ten Cate and A. Mileo (Eds.): RR 2015, LNCS 9209, pp. 45–60, 2015.
DOI: 10.1007/978-3-319-22002-4_5

domain-independent heuristics. In contrast, non-declarative heuristic approaches are very versatile [4,21,25] and provide ways to automatically generate heuristic planners from descriptions of a planning problem.
- The declarative approaches are typically still too complex, and this often defeats their stated goal of simplifying proofs of the different properties of the planning algorithms, such as completeness and termination.

In this paper, we will argue that *Transaction Logic* (or $\mathcal{TR}$) [8–10] provides multiple advantages for specifying and generalizing planning heuristics. Specifically, we will show that sophisticated planning heuristics, such as regression analysis, can be naturally represented in $\mathcal{TR}$ and that such representation can be used to express complex planning strategies such as *RSTRIPS*.

Transaction Logic is an extension of classical logic with dedicated support for specifying and reasoning about actions, including sequential and parallel execution, atomicity of transactions, and more. To illustrate the point, we take the regression analysis heuristic and a related planning algorithm *RSTRIPS* [23,28], and show that both the heuristics and the planning algorithm naturally lend themselves to compact representation in Transaction Logic. The resulting representation opens up new possibilities and, in particular, lets us prove the completeness of *RSTRIPS*. Clearly, existing formalizations of regression analysis and goal regression [24] are too complicated to be used as formal frameworks of such proofs. In contrast, the clear and compact form used to represent *RSTRIPS* in $\mathcal{TR}$ enables us to both implement and analyze *RSTRIPS* in a declarative framework, which is also much less complicated than its implementations even in Prolog [28] (which is not declarative).

The present paper continues the line of work in [2], where we (plain) *STRIPS* was represented in $\mathcal{TR}$ declaratively, extended, and made into a complete strategy. One can similarly apply $\mathcal{TR}$ to other heuristics proposed for planning algorithms [3]. As in [2], because *RSTRIPS* is cast here as a purely logical problem in a suitable general logic, a number of otherwise non-trivial extensions become easily achievable, and we get them almost for free. In particular, *RSTRIPS* planning can be naturally extended with derived predicates defined by rules. This endows the framework with the ability to express indirect effects of actions, but the resulting planning algorithm is still complete.

This paper is organized as follows. Section 2 reviews the *STRIPS* planning framework and extends it with the concept of regression of actions. Section 3 briefly overviews Transaction Logic in order to make this paper self-contained. Section 4 shows how regression of literals through actions can be computed. Section 5 shows how $\mathcal{TR}$ can represent *RSTRIPS* planning algorithm. Section 6 concludes the paper.

2   Extended *STRIPS*-Style Planning

In this section we first remind some standard concepts in logic and then introduce the *STRIPS* planning problem extended with the concept of regression.

We assume denumerable pairwise disjoint sets of variables $\mathcal{V}$, constants $\mathcal{C}$, extensional predicate symbols $\mathcal{P}_{ext}$, and intensional predicate symbols $\mathcal{P}_{int}$. As usual, atoms are formed by applying predicate symbols to ordered lists of constants or variables. Extending the logical signature with function symbols is straightforward in this framework, but we will not do it, as this is tangential to our aims. An atom is **extensional** if $p \in \mathcal{P}_{ext}$ and **intensional** if $p \in \mathcal{P}_{int}$. A **literal** is either an atom P or a negated extensional atom $\neg P$. Negated intensional atoms are not allowed (but such an extension is possible). In the original *STRIPS*, all predicates were extensional, and the addition of intentional predicates to *STRIPS* is a major enhancement, which allows us to deal with the so-called *ramification problem* [14], i.e., with indirect consequences of actions. Table 1 shows the syntax of our language.

Table 1. The syntax of the language for representing *STRIPS* planning problems.

Term	$t := V \mid c$	where $V \in \mathcal{V}, c \in \mathcal{C}$
Atom	$P_\tau := p(t_1, \ldots, t_k)$	where $p \in \mathcal{P}_\tau, \tau \in \{ext, int\}$
Literal	$L := P_{int} \mid P_{ext} \mid \neg P_{ext}$	
Rule	$R := P_{int} \leftarrow L_1 \wedge \cdots \wedge L_m$	where $m \geq 0$

Extensional predicates represent database facts: they can be directly manipulated (inserted or deleted) by actions. Intensional predicate symbols are used for atomic statements defined by *rules*—they are *not* affected by actions directly. Instead, actions make extensional facts true or false and this indirectly affects the dependent intensional atoms. These indirect effects are known as action *ramifications* in the literature.

A **fact** is a **ground** (i.e., variable-free) extensional atom. A set **S** of literals is **consistent** if there is no atom, *atm*, such that both *atm* and $\neg atm$ are in **S**.

A **rule** is a statement of the form *head* ← *body* where *head* is an intensional atom and body is a conjunction of literals. A **ground instance** of a rule, R, is any rule obtained from R by a substitution of variables with constants from $\mathcal{C}$ such that different occurrences of the same variable are always substituted with the same constant. Given a set **S** of literals and a ground rule of the form $atm \leftarrow \ell_1 \wedge \cdots \wedge \ell_m$, the rule is *true* in **S** if either $atm \in$ **S** or $\{\ell_1, \ldots, \ell_m\} \not\subseteq$ **S**. A (possibly non-ground) rule is *true* in **S** if all of its ground instances are true in **S**.

Definition 1 (State). *Given a set* $\mathbb{R}$ *of rules, a* **state** *is a consistent set* $\mathbf{S} = \mathbf{S}_{ext} \cup \mathbf{S}_{int}$ *of literals such that*

1. *For each fact atm, either* $atm \in \mathbf{S}_{ext}$ *or* $\neg atm \in \mathbf{S}_{ext}$.
2. *Every rule in* $\mathbb{R}$ *is true in* **S**.   □

Definition 2 (*STRIPS* **Action**). *A STRIPS **action** is a triple of the form*
$\alpha = \langle p_\alpha(X_1, ..., X_n), Pre_\alpha, E_\alpha \rangle$, *where*

- $p_\alpha(X_1, ..., X_n)$ *is an intensional atom in which* $X_1, ..., X_n$ *are variables and* $p_\alpha \in \mathcal{P}_{int}$ *is a predicate that is reserved to represent the action* α *and can be used for no other purpose;*
- Pre_α, *called the* **precondition** *of* α, *is a set that may include extensional as well as intensional literals;*
- E_α, *the* **effect** *of* α, *is a consistent set that may contain extensional literals only;*
- *The variables in* Pre_α *and* E_α *must occur in* $\{X_1, ..., X_n\}$.[1]   □

Note that the literals in Pre_α can be both extensional and intensional, while the literals in E_α can be extensional only.

Definition 3 (**Execution of a** *STRIPS* **Action**). *A STRIPS action* α *is* **executable** *in a state* **S** *if there is a substitution* $\theta : \mathcal{V} \longrightarrow \mathcal{C}$ *such that* $\theta(Pre_\alpha) \subseteq \mathbf{S}$. *A* **result of the execution** *of* α *with respect to* θ *is the state, denoted* $\theta(\alpha)(\mathbf{S})$, *defined as* $(\mathbf{S} \setminus \neg\theta(E_\alpha)) \cup \theta(E_\alpha)$, *where* $\neg E = \{\neg \ell \mid \ell \in E\}$. *In other words,* $\theta(\alpha)(\mathbf{S})$ *is* **S** *with all the effects of* $\theta(\alpha)$ *applied. When* α *is ground, we simply write* $\alpha(\mathbf{S})$.   □

Note that **S**′ is well-defined since $\theta(E_\alpha)$ is unique and consistent. Observe also that, if α has variables, the result of an execution, **S**′, depends on the chosen substitution θ.

The following simple example illustrates the above definition. We follow the standard logic programming convention whereby lowercase symbols represent constants and predicate symbols. The uppercase symbols denote variables that are implicitly universally quantified outside of the rules.

Example 1. Consider a world consisting of just two blocks and the action $pickup = \langle pickup(X, Y), \{clear(X)\}, \{\neg on(X, Y), clear(Y)\}\rangle$. Consider also the state $\mathbf{S} = \{clear(a), \neg clear(b), on(a, b), \neg on(b, a)\}$. Then the result of the execution of *pickup* at state **S** with respect to the substitution $\{X \to a, Y \to b\}$ is $\mathbf{S}' = \{clear(a), clear(b), \neg on(a, b), \neg on(b, a)\}$. It is also easy to see that *pickup* cannot be executed at **S** with respect to any substitution of the form $\{X \to b, Y \to ...\}$.   □

Definition 4 (**Planning Problem**). *A **planning problem** $\langle \mathbb{R}, \mathbb{A}, G, \mathbf{S} \rangle$ consists of a set of rules* $\mathbb{R}$, *a set of STRIPS actions* $\mathbb{A}$, *a set of literals* G, *called the* **goal** *of the planning problem, and an* **initial state** **S**. *A sequence of actions* $\sigma = \alpha_1, ..., \alpha_n$ *is a **planning solution** (or simply a **plan**) for the planning problem if:*

- $\alpha_1, ..., \alpha_n \in \mathbb{A}$; *and*

---

[1] Requiring all variables in Pre_α to occur in $\{X_1, ..., X_n\}$ is not essential: we can easily extend our framework and consider the extra variables to be existentially quantified.

– *there is a sequence of states* $\mathbf{S}_0, \mathbf{S}_1, \ldots, \mathbf{S}_n$ *such that*

- $\mathbf{S} = \mathbf{S}_0$ *and* $G \subseteq \mathbf{S}_n$ *(i.e., G is satisfied in the final state);*
- *for each* $0 < i \leq n$, α_i *is executable in state* $\mathbf{S}_{i-1}$ *and the result of that execution (for some substitution) is the state* $\mathbf{S}_i$.

In this case we will also say that $\mathbf{S}_0, \mathbf{S}_1, \ldots, \mathbf{S}_n$ *is an execution of* σ. □

Definition 5 (Non-redundant Plan). *Given a planing problem* $\langle \mathbb{R}, \mathbb{A}, G, \mathbf{S} \rangle$ *and a sequence of actions* $\sigma = \alpha_1, \ldots, \alpha_n$, *we call* σ *a* **non-redundant** *plan for* $\langle \mathbb{R}, \mathbb{A}, G, \mathbf{S} \rangle$ *if and only if:*

- σ *is a planning solution for* $\langle \mathbb{R}, \mathbb{A}, G, \mathbf{S} \rangle$;
- *None of* σ's *sub-sequences is a planning solution for the given planning problem.*

In other words, removing any action from σ *either makes the sequence non-executable at* $\mathbf{S}$ *or* G *is not satisfied after the execution.* □

In this section, we give a formal definition of the regression of literals through *STRIPS* actions. Section 4 shows how one can compute the regression of a literal through an action.

Definition 6 (Regression of a *STRIPS* Action). *Consider a STRIPS action* $\alpha = \langle p(\overline{X}), Pre, E \rangle$ *and a consistent set of fluents L. The* **regression of L through** α, *denoted* $\mathfrak{R}(\alpha, L)$,[2] *is a set of actions such that, for every* $\beta \in \mathfrak{R}(\alpha, L)$, $\beta = \langle p(\overline{X}), Pre_\beta, E \rangle$, *where* $Pre_\beta \supseteq Pre$ *is a minimal (with respect to $\subsetneq$) set of fluents satisfying the following condition: For every state* $\mathbf{S}$ *and substitution* θ *such that* $\theta(\alpha)(\mathbf{S})$ *exists, if* $\mathbf{S} \models \theta(Pre_\beta) \wedge \theta(L)$, *then* $\theta(\alpha)(\mathbf{S}) \models \theta(L)$. *In other words, β has the same effects as α, but its precondition is more restrictive and it preserves (does not destroy) the set of literals L.*

Each action in $\mathfrak{R}(\alpha, L)$ *will also be called a regression of L via α.* □

The *minimal set of fluents* in this definition is, as noted, with respect to subset, i.e., there is no action $\beta' = \langle p(\overline{X}), Pre_{\beta'}, E \rangle$ such that $Pre_{\beta'} \subsetneq Pre_\beta$ and β' satisfies the conditions of Definition 6.

Consider $\beta \in \mathfrak{R}(\alpha, L)$ and let $\check{\beta} = \langle p(\overline{X}), Pre_\beta \cup L, E \rangle$. We will call $\check{\beta}$ a **restricted** regression of L through α and denote the set of such actions by $\check{\mathfrak{R}}(\alpha, L)$. We will mostly use the restricted regressions of actions in the representation of *RSTRIPS* planning algorithm.

3   Overview of Transaction Logic

To make this paper self-contained, we provide a brief introduction to the relevant *subset* of $\mathcal{TR}$ [5,7–10] needed for the understanding of this paper.

---

[2] We simply write $\mathfrak{R}(\alpha, \ell)$ whenever L just contains a single literal ℓ.

As an extension of first-order predicate calculus, $\mathcal{TR}$ shares much of its syntax with that calculus. One of the new connectives that $\mathcal{TR}$ adds to the calculus is the **serial conjunction**, denoted $\otimes$. It is binary associative, and non-commutative. The formula $\phi \otimes \psi$ represents a composite action of *execution* of ϕ followed by an execution of ψ. When ϕ and ψ are regular first-order formulas, $\phi \otimes \psi$ reduces to the usual first-order conjunction, $\phi \wedge \psi$. The logic also introduces other connectives to support hypothetical reasoning, concurrent execution, etc., but these are not going to be used here.

To take the *frame problem* out of many considerations in $\mathcal{TR}$, it has an extensible mechanism of **elementary updates** (see [6,7,9,10]). Due to the definition of *STRIPS* actions, we just need the following two types of elementary updates (actions): $+p(t_1, \ldots, t_n)$ and $-p(t_1, \ldots, t_n)$, where $p(t_1, \ldots, t_n)$ denotes an *extensional* atom. Given a state **S** and a *ground* elementary action $+p(a_1, \ldots, a_n)$, an execution of $+p(a_1, \ldots, a_n)$ at state **S** deletes the literal $\neg p(a_1, \ldots, a_n)$ and adds the literal $p(a_1, \ldots, a_n)$. Similarly, executing $-p(a_1, \ldots, a_n)$ results in a state that is exactly like **S**, but $p(a_1, \ldots, a_n)$ is deleted and $\neg p(a_1, \ldots, a_n)$ is added. If $p(a_1, \ldots, a_n) \in$ **S**, the action $+p(a_1, \ldots, a_n)$ has no effect, and similarly for $-p(a_1, \ldots, a_n)$.

We define **complex actions** using **serial rules**, which are statements of the form

$$h \leftarrow b_1 \otimes b_2 \otimes \ldots \otimes b_n. \qquad (1)$$

where h is an atomic formula denoting the complex action and b_1, ..., b_n are literals or elementary actions. This means that h is a complex action and one way to execute h is to execute b_1 then b_2, etc., and finally to execute b_n. Note that we have regular first-order as well as serial-Horn rules. For simplicity, we assume that the sets of intentional predicates that can appear in the heads of regular rules and those in the heads of serial rules are disjoint. *Extensional atoms* and *intentional atoms* that can appear in the states (see Definition 1) will be called **fluents**. Note that a serial rule all of whose body literals are fluents is essentially a regular rule, since all the $\otimes$-connectives can be replaced with $\wedge$. Therefore, one can view the regular rules as a special case of serial rules.

The following example illustrates the above concepts where we continue to use the standard logic programming convention regarding capitalization of variables, which are assumed to be universally quantified outside of the rules. It is common practice to omit quantifiers.

$$
\begin{aligned}
move(X, Y) \;\leftarrow\; & (on(X, Z) \wedge clear(X) \\
& \wedge clear(Y) \wedge \neg tooHeavy(X)) \otimes \\
& -on(X, Z) \otimes +on(X, Y) \otimes \\
& -clear(Y). \\
tooHeavy(X) \;\leftarrow\; & weight(X, W) \wedge limit(L) \wedge \\
& W < L. \\
? - \; & move(blk1, blk15) \otimes move(SomeBlk, blk1).
\end{aligned}
$$

Here *on*, *clear*, *tooHeavy*, *weight*, etc., are fluents and *move* is an action. The predicate *tooHeavy* is an intentional fluent, while *on*, *clear*, and *weight* are extensional fluents. The actions $+on(...)$, $-clear(...)$, and $-on(...)$ are elementary and the intentional predicate *move* is a complex action. This example illustrates several features of Transaction Logic. The first rule is a serial rule defining a complex action of moving a block from one place to another. The second rule defines the intensional fluent *tooHeavy*, which is used in the definition of *move* (under the scope of default negation). As the second rule does not include any action, it is a regular rule.

The last statement above is a *request to execute* a composite action, which is analogous to a query in logic programming. The request is to move block *blk1* from its current position to the top of *blk15* and then find some other block and move it on top of *blk1*. Traditional logic programming offers no logical semantics for updates, so if after placing *blk1* on top of *blk15* the second operation ($move(SomeBlk, blk1)$) fails (say, all available blocks are too heavy), the effects of the first operation will persist and the underlying database becomes corrupted. In contrast, Transaction Logic gives update operators a logical semantics of an *atomic database transaction*. This means that if any part of the transaction fails, the effect is as if nothing was done at all. For example, if the second action in our example fails, all actions are "backtracked over" and the underlying database state remains unchanged.

$\mathcal{TR}$'s semantics is given in purely model-theoretic terms and here we will just give an informal overview. The truth of any action in $\mathcal{TR}$ is determined over sequences of states—**execution paths**—which makes it possible to think of truth assignments in $\mathcal{TR}$'s models as executions. If an action, ψ, defined by a set of serial rules, $\mathbb{P}$, evaluates to true over a sequence of states $\mathbf{D}_0, \ldots, \mathbf{D}_n$, we say that it can *execute* at state $\mathbf{D}_0$ by passing through the states $\mathbf{D}_1$, ..., $\mathbf{D}_{n-1}$, ending in the final state $\mathbf{D}_n$. This is captured by the notion of **executional entailment**, which is written as follows:

$$\mathbb{P}, \mathbf{D}_0 \ldots \mathbf{D}_n \models \psi \tag{2}$$

Due to lack of space, we put more examples about $\mathcal{TR}$ in the full report.[3]

Various inference systems for serial-Horn $\mathcal{TR}$ [7] are similar to the well-known SLD resolution proof strategy for Horn clauses plus some $\mathcal{TR}$-specific inference rules and axioms. Given a set of serial rules, $\mathbb{P}$, and a *serial goal*, ψ (i.e., a formula that has the form of a body of a serial rule such as (1)), these inference systems prove statements of the form $\mathbb{P}, \mathbf{D} \cdots \vdash \psi$, called **sequents**. A proof of a sequent of this form is interpreted as a proof that action ψ defined by the rules in $\mathbb{P}$ can be successfully executed starting at state $\mathbf{D}$.

An inference succeeds iff it finds an execution for the transaction ψ. The execution is a sequence of database states $\mathbf{D}_1, \ldots, \mathbf{D}_n$ such that $\mathbb{P}, \mathbf{D}\,\mathbf{D}_1 \ldots \mathbf{D}_n \models \psi$. We will use the following inference system in our planning application. For simplicity, we present only a version for ground facts and rules. The inference rules can be read either top-to-bottom (if *top* is proved then *bottom* is proved) or bottom-to-top (to prove *bottom* one needs to prove *top*).

---

[3] http://ewl.cewit.stonybrook.edu/planning/RSTRIPS-TR-full.pdf.

Definition 7 ($\mathcal{TR}$ Inference System). *Let $\mathbb{P}$ be a set of rules (serial or regular) and* $\mathbf{D}$, $\mathbf{D}_1$, $\mathbf{D}_2$ *denote states.*

– *Axiom:* $\mathbb{P}, \mathbf{D} \cdots \vdash ()$, *where () is an empty clause (which is true at every state).*
– *Inference Rules*
   1. *Applying transaction definition: Suppose $t \leftarrow body$ is a rule in* $\mathbb{P}$.

$$\frac{\mathbb{P}, \mathbf{D} \cdots \vdash body \otimes rest}{\mathbb{P}, \mathbf{D} \cdots \vdash t \otimes rest} \tag{3}$$

   2. *Querying the database: If* $\mathbf{D} \models t$ *then*

$$\frac{\mathbb{P}, \mathbf{D} \cdots \vdash rest}{\mathbb{P}, \mathbf{D} \cdots \vdash t \otimes rest} \tag{4}$$

   3. *Performing elementary updates: If the elementary update t changes the state $\mathbf{D}_1$ into the state $\mathbf{D}_2$ then*

$$\frac{\mathbb{P}, \mathbf{D}_2 \cdots \vdash rest}{\mathbb{P}, \mathbf{D}_1 \cdots \vdash t \otimes rest} \tag{5}$$

A **proof** of a sequent, seq_n, is a series of sequents, seq_1, seq_2, ..., seq_{n-1}, seq_n, where each seq_i is either an axiom-sequent or is derived from earlier sequents by one of the above inference rules. This inference system has been proven sound and complete with respect to the model theory of $\mathcal{TR}$ [7]. This means that if ϕ is a serial goal, the executional entailment $\mathbb{P}, \mathbf{D}_0 \mathbf{D}_1 \ldots \mathbf{D}_n \models \phi$ holds if and only if there is a proof of $\mathbb{P}, \mathbf{D}_0 \cdots \vdash \phi$ over the execution path $\mathbf{D}_0, \mathbf{D}_1, \ldots, \mathbf{D}_n$. In this case, we will also say that such a proof derives $\mathbb{P}, \mathbf{D}_0 \mathbf{D}_1 \ldots \mathbf{D}_n \vdash \phi$.

4   Computation of Regression

In this section, we briefly explain how the previously introduced regression of actions (Definition 6) can be computed. This computation is a key component of the *RSTRIPS* planning algorithm.

In the following we will be using an *identity* operator, $=$, which will be treated as an immutable extensional predicate, i.e., a predicate defined by a non-changeable set of facts: For any state $\mathbf{S}$ and a pair of constants or ground fluents ℓ and ℓ', $(\ell = \ell') \in \mathbf{S}$ if and only if ℓ and ℓ are identical. Similarly, *non-identity* is defined as follows: $\ell \neq \ell' \in \mathbf{S}$ if and only if ℓ, ℓ' are distinct.

To illustrate regression, consider a *STRIPS* action *copy* $=$ $\langle copy(Src, Dest, V), \{value(Src, V)\}, \{\neg value(Dest, V'), value(Dest, V)\} \rangle$ from the *Register Exchange* example in [2]. For convenience, this example is also found in the full report[4] along with *RSTRIPS* planning rules. Here, *copy* $\in$

---
[4] http://ewl.cewit.stonybrook.edu/planning/RSTRIPS-TR-full.pdf.

$\mathfrak{R}(copy, value(Src, V))$ since for every state $\mathbf{S}$ and substitution θ such that $\theta(copy)(\mathbf{S})$ exists, if $\mathbf{S} \models \theta(value(Src, V))$, then $\theta(\alpha)(\mathbf{S}) \models \theta(value(Src, V))$. This example is a special case of the following property of regression, which directly follows from the definitions: if ℓ is an extensional literal and $\alpha = \langle p(\overline{X}), Pre, E \rangle$ is a STRIPS action,

- $\mathfrak{R}(\alpha, \ell) = \emptyset$ if and only if, for every ground substitution, $\neg\theta(\ell) \in \theta(E)$.
- $\mathfrak{R}(\alpha, \ell) = \{\alpha\}$ if and only if, for every ground substitution $\neg\theta(\ell) \notin \theta(E)$.

The following proposition and lemmas present a method to compute regression. The method is complete for extensional literals; for intentional literals, it yields some, but not always all, regressions.

Proposition 1 (Regression of Sets of Literals). *Given a set of literals $L = L_1 \cup L_2$ and a STRIPS action $\alpha = \langle p(\overline{X}), Pre_\alpha, E_\alpha \rangle$, let $\beta_1 \in \mathfrak{R}(\alpha, L_1)$ and $\beta_2 \in \mathfrak{R}(\alpha, L_2)$, where $\beta_1 = \langle p(\overline{X}), Pre_{\beta_1}, E_\alpha \rangle$ and $\beta_2 = \langle p(\overline{X}), Pre_{\beta_2}, E_\alpha \rangle$. There is some $\beta = \langle p(\overline{X}), Pre_\beta, E_\alpha \rangle$ such that $Pre_\beta \subseteq Pre_{\beta_1} \cup Pre_{\beta_2}$ and $\beta \in \mathfrak{R}(\alpha, L)$.*

Proof. From the assumptions, it follows that for every state $\mathbf{S}$ and substitution θ such that $\theta(\alpha)(\mathbf{S})$ exists, if $\mathbf{S} \models \theta(Pre_{\beta_1} \cup Pre_{\beta_2}) \wedge \theta(L)$, then $\theta(\alpha)(\mathbf{S}) \models \theta(L)$.

To find a minimal subset of $Pre_{\beta_1} \cup Pre_{\beta_2}$ satisfying the regression property, one can repeatedly remove elements from $Pre_{\beta_1} \cup Pre_{\beta_2}$ and check if the regression property still holds. When no removable elements remain, we get a desired set Pre_β. □

Lemma 1. (Regression of Extensional Literals). *Consider an extensional literal ℓ and a STRIPS action $\alpha = \langle p(\overline{X}), Pre, E \rangle$ where α and ℓ do not share variables. Let $Pre_\beta = Pre \cup \{ \ell \neq e \mid \neg e \in E \ \wedge \ \exists \theta \ s.t. \ \theta(e) = \theta(\ell) \}$. Then $\beta \in \mathfrak{R}(\alpha, L)$, where $\beta = \langle p(\overline{X}), Pre_\beta, E \rangle$.*

Proof. Let $\mathbf{S}$ be a state and there is θ such that $\theta(\alpha)(\mathbf{S})$ exists. Clearly, if $\mathbf{S} \models \theta(Pre_\beta)$, there is no $\neg e \in E$ such that $\theta(e) = \theta(\ell)$. Therefore, if $\mathbf{S} \models \theta(Pre_\beta) \wedge \theta(\ell)$, then $\theta(\alpha)(\mathbf{S}) \models \ell$.

We need to show that Pre_β is a minimal set of literals satisfying the above property. Assume, to the contrary, that there is some $Pre_{\beta'}$, $Pre \subseteq Pre_{\beta'} \subsetneq Pre_\beta$, such that for every state $\mathbf{S}$ and substitution θ, if $\theta(\alpha)(\mathbf{S})$ exists and $\mathbf{S} \models \theta(Pre_{\beta'}) \wedge \theta(\ell)$, then $\theta(\alpha)(\mathbf{S}) \models \theta(\ell)$. Since $Pre_{\beta'} \subset Pre_\beta$, there must be $(\ell \neq e) \in Pre_\beta \setminus Pre_{\beta'}$. Let θ_1 be a substitution such that $\theta_1(e) = \theta_1(\ell)$. In that case for every $\mathbf{S}$ such that $\theta_1(\alpha)(\mathbf{S})$ exists, $\mathbf{S} \models \theta_1(Pre_{\beta'}) \wedge \theta_1(\ell)$ but $\theta_1(\alpha)(\mathbf{S}) \not\models \theta(\ell)$, since $\theta_1(\neg\ell) = \theta(\neg e) \in \theta_1(E)$. This contradicts the assumption that $\theta_1(\ell) \in \mathbf{S}$. Thus Pre_β is a minimal set of fluents satisfying the regression condition for ℓ, so $\beta \in \mathfrak{R}(\alpha, \ell)$. □

To illustrate the lemma, consider an extensional literal $value(R, V'')$ and the STRIPS action $copy = \langle copy(S, D, V), Pre_{copy}, E_{cppy} \rangle$, where $Pre_{copy} = \{value(S, V)\}$ and $E_{copy} = \{\neg value(D, V'), value(D, V)\}$. Then $\beta \in \mathfrak{R}(copy, value(R, V''))$, where $\beta = \langle copy_\beta(S, D, V), Pre_\beta, E_{copy} \rangle$ and $Pre_\beta = Pre_{copy} \cup \{value(R, V'') \neq value(D, V')\}$.

Lemma 2 (Regression of Intensional Literals). *Consider a set of rules* $\mathbb{R}$, *an intensional literal* ℓ, *and a STRIPS action* $\alpha = \langle p(\overline{X}), Pre, E \rangle$, *where* α *and* ℓ *do not share variables. Let* L *be a minimal set of extensional literals such that* $\mathbb{R} \cup L \cup \{\leftarrow \ell\}$ *has an SLD-refutation [19]. Then for every* $\beta \in \mathfrak{R}(\alpha, L)$ *of the form* $\beta = \langle p(\overline{X}), Pre_\beta, E \rangle$, *there is* $L_\beta \subseteq Pre_\beta \cup L$ *such that* $\langle p(\overline{X}), L_\beta, E \rangle \in \mathfrak{R}(\alpha, \ell)$.

Proof. By Definition 6, for every state $\mathbf{S}$ and substitution θ such that $\theta(\alpha)(\mathbf{S})$ exists, if $\mathbf{S} \models \theta(Pre_\beta) \wedge \theta(L)$, then $\theta(\alpha)(\mathbf{S}) \models \theta(L)$. Due to the soundness of SLD-refutation [19], if $\mathbf{S} \models \theta(Pre_\beta) \wedge \theta(L)$ then $\mathbf{S} \models \theta(\ell)$; and if $\theta(\alpha)(\mathbf{S}) \models \theta(L)$ then $\theta(\alpha)(\mathbf{S}) \models \theta(\ell)$. Therefore, for every state $\mathbf{S}$ and substitution θ such that $\theta(\alpha)(\mathbf{S})$ exists, if $\mathbf{S} \models \theta(Pre_\beta) \wedge \theta(L) \wedge \theta(\ell)$, then $\theta(\alpha)(\mathbf{S}) \models \theta(\ell)$. Therefore, $Pre_\beta \cup L$ satisfies the conditions for regressing ℓ through α except, possibly, minimality. To get the minimality, we can start removing elements from this set, as in Proposition 1, until a minimal set is reached. $\qquad\square$

Definition 8. (Regression Deterministic Action). *A STRIPS action* $\alpha = \langle p(\overline{X}), Pre_\alpha, E \rangle$ *is called* ***regression-deterministic*** *if for every set of literals* L, *one of the following holds:*

- *There exists* $\beta \in \mathfrak{R}(\alpha, L)$ *such that* $Pre_\beta \setminus Pre_\alpha$ *is a set of literals of the form* $\ell = e$ *or* $\ell \neq e$.
- $\mathfrak{R}(\alpha, L) = \emptyset$.

Similarly, a set of actions $\mathbb{A}$ *is* ***regression-deterministic*** *if all of its actions are regression-deterministic.* $\qquad\square$

Clearly, if a set of actions $\mathbb{A}$ is regression-deterministic, one can find an action $\beta \in \mathfrak{R}(\alpha, L)$ for every $\alpha \in \mathbb{A}$ and set of literals L using Lemmas 1 and 2. From now on, we assume that the set of actions $\mathbb{A}$ is regression-deterministic and is closed under regression and restricted regression.

5   The *RSTRIPS* Planner

The idea of using $\mathcal{TR}$ as a planning formalism and an encoding of *STRIPS* as a set of $\mathcal{TR}$ rules first appeared informally in the unpublished report [7]. The encoding did not include ramification and intensional predicates. Based on that encoding, we proposed a *non-linear* and *complete* planning algorithm in [2]. In this paper, we extend the original method with regression analysis and use $\mathcal{TR}$ to represent the *RSTRIPS* planning algorithm. This also generalizes the original *RSTRIPS* with intentional predicates and we prove the completeness of the resulting planner. To the best of our knowledge, completeness of *RSTRIPS* has not been proven before.

Regression analysis of literals, as a search heuristic for planners, can be used to improve the performance of planning strategies. The idea behind planning with regression is that the already achieved goals should be protected so that subsequent actions of the planner would not "unachieve" those goals [27].

To keep our encoding simple, we assume a built-in 3-ary predicate $regress$ such that, for any state $\mathbf{S}$, $regress(L, p_\alpha(\overline{X}), p_{\alpha'}(\overline{X}))$ is true (on any path) if and only if $\alpha' \in (\mathfrak{R}(\alpha, L) \cup \breve{\mathfrak{R}}(\alpha, L) \cup \{\alpha\})$. During plan construction, $RSTRIPS$ may consider a subgoal that cannot be achieved without unachieving an already achieved goal [28]. Instead of checking for unachieved goals *after* performing actions (and undoing these actions when an unachieved goal is found), $RSTRIPS$ verifies that no unachieving will take place *before* performing each action by modifying action preconditions using regression.

Definition 9 (Enforcement Operator). *Let G be a set of extensional literals. We define $Enf(G) = \{+p \mid p \in G\} \cup \{-p \mid \neg p \in G\}$. In other words, $Enf(G)$ is the set of elementary updates that makes G true.*   □

Next we introduce a natural correspondence between $STRIPS$ actions and $T\mathcal{R}$ rules.

Definition 10 (Actions as $T\mathcal{R}$ Rules). *Let $\alpha = \langle p_\alpha(\overline{X}), Pre_\alpha, E_\alpha \rangle$ be a STRIPS action. We define its **corresponding $T\mathcal{R}$ rule**, $tr(\alpha)$, to be a rule of the form*

$$p_\alpha(\overline{X}) \leftarrow (\wedge_{\ell \in Pre_\alpha} \ell) \otimes (\otimes_{u \in Enf(E_\alpha)} u). \tag{6}$$

Note that in (6) the actual order of action execution in the last component, $\otimes_{u \in Enf(E_\alpha)} u$, is immaterial, since all such executions happen to lead to the same state.

We now define a set of $T\mathcal{R}$ clauses that simulate the well-known $RSTRIPS$ planning algorithm and extend this algorithm to handle intentional predicates and rules. The reader familiar with the $RSTRIPS$ planner may notice that these rules are essentially a natural, more concise, and more general verbalization of the classical $RSTRIPS$ algorithm [12]. These rules constitute a *complete* planner when evaluated with the $T\mathcal{R}$ proof theory.

Definition 11 ($T\mathcal{R}$ Planning Rules with Regression Analysis). *Let $\Pi = \langle \mathbb{R}, \mathbb{A}, G, \mathbf{S} \rangle$ be a STRIPS planning problem (see Definition 4). We define a set of $T\mathcal{R}$ rules, $\mathbb{P}^r(\Pi)$, which simulates the RSTRIPS planning algorithm. $\mathbb{P}^r(\Pi)$ has three disjoint parts: $\mathbb{P}^r_{\mathbb{R}}$, $\mathbb{P}^r_{\mathbb{A}}$, and $\mathbb{P}^r_G$, as described below.*

- *The $\mathbb{P}^r_{\mathbb{R}}$ part: for each rule $p(\overline{X}) \leftarrow p_1(\overline{X}_1) \wedge \cdots \wedge p_n(\overline{X}_n)$ in $\mathbb{R}$, $\mathbb{P}^r_{\mathbb{R}}$ is a set of rules of the following form—one rule per permutation $\langle i_1, \ldots, i_n \rangle$:*

$$\begin{aligned} achieve^r_p(\overline{X}, L) \leftarrow\ & achieve^r_{p_{i_1}}(\overline{X}_{i_1}, L) \otimes \\ & achieve^r_{p_{i_2}}(\overline{X}_{i_2}, L \cup \{p_{i_1}(\overline{X}_{i_1})\}) \otimes \ldots \otimes \\ & achieve^r_{p_{i_n}}(\overline{X}_{i_n}, L \cup \{p_{i_1}(\overline{X}_{i_1}), \ldots, p_{i_{n-1}}(\overline{X}_{i_{n-1}})\}) \end{aligned} \tag{P1}$$

Rule (P1) extends the classical RSTRIPS setting with intentional predicates and ramification of actions.

– *The part* $\mathbb{P}_{\mathbb{A}}^r = \mathbb{P}_{actions}^r \cup \mathbb{P}_{atoms}^r \cup \mathbb{P}_{achieves}^r$ *is constructed out of the actions in $\mathbb{A}$ as follows:*

• $\mathbb{P}_{actions}^r$: *for each $\alpha \in \mathbb{A}$, $\mathbb{P}_{actions}^r$ has a rule of the form*

$$p_\alpha(\overline{X}) \leftarrow (\wedge_{\ell \in Pre_\alpha} \ell) \otimes (\otimes_{u \in Enf(E_\alpha)} u). \tag{P2}$$

This is the $\mathcal{TR}$ rule that corresponds to the action α, introduced in Definition 10.

• $\mathbb{P}_{atoms}^r = \mathbb{P}_{achieved}^r \cup \mathbb{P}_{enforced}^r$ *has two disjoint parts as follows:*

– $\mathbb{P}_{achieved}^r$: *for each extensional predicate $p \in \mathcal{P}_{ext}$, $\mathbb{P}_{achieved}^r$ has the rules*

$$\begin{aligned} achieve_p^r(\overline{X}, L) &\leftarrow p(\overline{X}). \\ achieve_{not\_p}^r(\overline{X}, L) &\leftarrow \neg p(\overline{X}). \end{aligned} \tag{P3}$$

These rules say that if an extensional literal is true in a state then that literal has already been achieved as a goal.

– $\mathbb{P}_{enforced}^r$: *for each action $\alpha = \langle p_\alpha(\overline{X}), Pre_\alpha, E_\alpha \rangle$ in $\mathbb{A}$ and each $e(\overline{Y}) \in E_\alpha$, $\mathbb{P}_{enforced}^r$ has the following rule:*

$$\begin{aligned} achieve_e^r(\overline{Y}, L) &\leftarrow regress(L, p_\alpha(\overline{X}), p_{\alpha'}(\overline{X})) \\ &\otimes execute_{p_{\alpha'}}^r(\overline{X}, L). \end{aligned} \tag{P4}$$

This rule says that one way to achieve a goal that occurs in the effects of an action is to execute that action after regressing the "protected" literals L through that action.

• $\mathbb{P}_{achieves}^r$: *for each action $\alpha = \langle p_\alpha(\overline{X}), Pre_\alpha, E_\alpha \rangle$ in $\mathbb{A}$ where $Pre_\alpha = \{p_1(\overline{X}_1), \ldots, p_n(\overline{X}_n)\}$, $\mathbb{P}_{achieves}^r$ is a set of rules of the following form, one per permutation $\langle i_1, \ldots, i_n \rangle$:*

$$\begin{aligned} execute_{p_\alpha}^r(\overline{X}, L) &\leftarrow achieve_{p_{i_1}}^r(\overline{X}_{i_1}, L) \otimes s \\ &\quad achieve_{p_{i_2}}^r(\overline{X}_{i_2}, L \cup \{p_{i_1}(\overline{X}_{i_1})\}) \otimes \ldots \otimes \\ &\quad achieve_{p_{i_n}}^r(\overline{X}_{i_n}, L \cup \{p_{i_1}(\overline{X}_{i_1}), \ldots, p_{i_{n-1}}(\overline{X}_{i_{n-1}})\}) \\ &\quad \otimes p_\alpha(\overline{X}). \end{aligned} \tag{P5}$$

This means that to execute an action, one must first achieve the precondition of the action while making sure to not unachieve the already achieved parts of the precondition.

– $\mathbb{P}_G^r$: *Let $G = \{g_1, \ldots, g_k\}$. Then $\mathbb{P}_G^r$ is a set of the following rules, one per permutation $\langle i_1, \ldots, i_n \rangle$:*

$$\begin{aligned} achieve_G &\leftarrow achieve_{g_{i_1}}^r(L) \\ &\quad \otimes achieve_{g_{i_2}}^r(L \cup \{g_{i_1}\}) \otimes \ldots \\ &\quad \otimes achieve_{g_{i_n}}^r(L \cup \{g_{i_1}, \ldots, g_{i_{n-1}}\}). \end{aligned} \tag{P6}$$

Due to space limitation, we cannot include an example of $\mathcal{TR}$-based $RSTRIPS$ planning here. Instead, we refer the reader to Example 3 in the full report.[5]

Given a set $\mathbb{R}$ of rules, a set $\mathbb{A}$ of $STRIPS$ actions, an initial state $\mathbf{S}$, and a goal G, Definition 11 gives a set of $\mathcal{TR}$ rules that specify a planning strategy for that problem. To find a solution for that planning problem, one simply needs to place the request

$$? - achieve_G . \tag{7}$$

at a desired initial state and use the $\mathcal{TR}$'s inference system of Sect. 3 to find a proof. The inference system in question is sound and complete for *serial clauses* [5,7,9], and the rules in Definition 11 satisfy that requirement.

As mentioned before, a solution plan for a $STRIPS$ planning problem is a sequence of actions leading to a state that satisfies the planning goal. Such a sequence can be extracted by picking out the atoms of the form p_α from a successful derivation branch generated by the $\mathcal{TR}$ inference system. Since each p_α uniquely corresponds to a $STRIPS$ action, this provides us with the requisite sequence of actions that forms a plan.

Suppose $seq_0, \ldots, seq_m$ is a deduction by the $\mathcal{TR}$ inference system. Let $i_1, \ldots, i_n$ be exactly those indexes in that deduction where the inference rule (3) was applied to some sequent using a rule of the form $tr(\alpha_{i_r})$ introduced in Definition 10. We will call $\alpha_{i_1}, \ldots, \alpha_{i_n}$ the ***pivoting sequence of actions***. The corresponding ***pivoting sequence of states*** $\mathbf{D}_{i_1}, \ldots, \mathbf{D}_{i_n}$ is a sequence where each $\mathbf{D}_{i_r}$, $1 \leq r \leq n$, is the state at which α_{i_r} is applied. We will prove that the pivoting sequence of actions is a solution to the planning problem. The proofs are found in the full report.

Theorem 1 (Soundness of $\mathcal{TR}$ Planning Solutions). *Consider a STRIPS planning problem* $\Pi = \langle \mathbb{R}, \mathbb{A}, G, \mathbf{D}_0 \rangle$ *and let* $\mathbb{P}^r(\Pi)$ *be the corresponding set of* $\mathcal{TR}$ *rules, as in Definition 11. Then any pivoting sequence of actions in the derivation of the sequent* $\mathbb{P}^r(\Pi), \mathbf{D}_0 \ldots \mathbf{D}_m \vdash achieve_G$ *is a solution plan.*

Completeness of a planning strategy means that, for any $STRIPS$ planning problem, if there is a solution, the planner will find *at least one* plan.

Theorem 2 (Completeness of $\mathcal{TR}$ Planning). *Given a STRIPS planning problem* $\Pi = \langle \mathbb{R}, \mathbb{A}, G, \mathbf{D}_0 \rangle$, *let* $\mathbb{P}^r(\Pi)$ *be the corresponding set of* $\mathcal{TR}$ *rules as in Definition 11. If there is a plan for* Π *then* $\mathcal{TR}$ *inference system will fins some plan using the rules* $\mathbb{P}^r(\Pi)$, *as described in Definition 11.*

Theorem 3 (Decidability). *$\mathcal{TR}$-based planners for RSTRIPS always terminates.*

6   Conclusion

This paper has demonstrated that the use of Transaction Logic accrues significant benefits in the area of planning. As an illustration, we have shown that

---

[5] http://ewl.cewit.stonybrook.edu/planning/RSTRIPS-TR-full.pdf.

sophisticated planning heuristics and algorithms, such as regression analysis and *RSTRIPS*, can be naturally represented in $\mathcal{TR}$ and that the use of this powerful logic opens up new possibilities for generalizations and proving properties of planning algorithms. For instance, just by using this logic, we were able to extend *RSTRIPS* with action ramification almost for free. Furthermore, benefiting from the proof theory, we were able to establish the completeness and termination of the resulting strategy. In the full report,[6] we also present our experiments that show that *RSTRIPS* can be orders of magnitude better than *STRIPS* both in time and space. These non-trivial insights were acquired merely due to the use of $\mathcal{TR}$ and not much else. The same technique can be used to cast even more advanced strategies such as GraphPlan, *ABSTRIPS* [23], and HTN [20] as $\mathcal{TR}$ rules.

There are several promising directions to continue this work. One is to investigate other planning strategies and, hopefully, accrue similar benefits. Other possible directions include non-linear plans and plans with loops [16,17,30]. For instance non-linear plans could be represented using Concurrent Transaction Logic [8], while loops are easily representable using recursive actions in $\mathcal{TR}$.

Acknowledgments. We are thankful to the anonymous referees for their thorough reviews and suggestions.

References

1. Bacchus, F., Kabanza, F.: Using temporal logics to express search control knowledge for planning. Artif. Intell. **116**(12), 123–191 (2000). http://www.sciencedirect.com/science/article/pii/S0004370299000715
2. Basseda, R., Kifer, M., Bonner, A.J.: Planning with transaction logic. In: Kontchakov, R., Mugnier, M.-L. (eds.) RR 2014. LNCS, vol. 8741, pp. 29–44. Springer, Heidelberg (2014)
3. Bonet, B., van den Briel, M.: Flow-based heuristics for optimal planning: landmarks and merges. In: Chien, S., Do, M.B., Fern, A., Ruml, W. (eds.) Proceedings of the Twenty Fourth International Conference on Automated Planning and Scheduling, ICAPS 2014. AAAI, Portsmouth (2014). http://www.aaai.org/ocs/index.php/ICAPS/ICAPS14/paper/view/7933
4. Bonet, B., Geffner, H.: Planning as heuristic search. Artif. Intell. **129**(1–2), 5–33 (2001). http://dx.doi.org/10.1016/S0004-3702(01)00108-4
5. Bonner, A., Kifer, M.: Transaction logic programming. In: International Conference on Logic Programming, pp. 257–282. MIT Press, Budaspest (1993)
6. Bonner, A., Kifer, M.: Applications of transaction logic to knowledgerepresentation. In: Gabbay, D.M., Ohlbach, H.J. (eds.) ICTL 1994. LNCS, vol. 827, pp. 67–81. Springer, Heidelberg (1994)
7. Bonner, A., Kifer, M.: Transaction logic programming (or a logic of declarative and procedural knowledge). Technical report CSRI-323, University of Toronto (November 1995). http://www.cs.toronto.edu/~bonner/transaction-logic.html

---

[6] http://ewl.cewit.stonybrook.edu/planning/RSTRIPS-TR-full.pdf.

8. Bonner, A., Kifer, M.: Concurrency and communication in transaction logic. In: Joint International Conference and Symposium on Logic Programming, pp. 142–156. MIT Press, Bonn, September 1996

9. Bonner, A., Kifer, M.: A logic for programming database transactions. In: Chomicki, J., Saake, G. (eds.) Logics for Databases and Information Systems, pp. 117–166. Kluwer Academic Publishers, Norwell (1998). Chap. 5

10. Bonner, A.J., Kifer, M.: An overview of transaction logic. Theo. Comput. Sci. **133**, 205–265 (1994)

11. Doherty, P., Kvarnström, J., Heintz, F.: A temporal logic-based planning and execution monitoring framework for unmanned aircraft systems. Auton. Agent. Multi-Agent Syst. **19**(3), 332–377 (2009). http://dx.doi.org/10.1007/s10458-009-9079-8

12. Fikes, R.E., Nilsson, N.J.: STRIPS: A new approach to the application of theorem proving to problem solving. Artif. Intell. **2**(34), 189–208 (1971)

13. Gerevini, A., Schubert, L.: Accelerating partial-order planners: some techniques for effective search control and pruning. J. Artif. Intell. Res. (JAIR) **5**, 95–137 (1996)

14. Giunchiglia, E., Lifschitz, V.: Dependent fluents. In: Proceedings of International Joint Conference on Artificial Intelligence (IJCAI), pp. 1964–1969 (1995)

15. Joslin, D., Pollack, M.E.: Least-cost flaw repair: A plan refinement strategy for partial-order planning. In: Proceedings of the Twelth National Conference on Artificial Intelligence, vol. 2, pp. 1004–1009. American Association for Artificial Intelligence, AAAI 1994, Menlo Park (1994). http://dl.acm.org/citation.cfm?id=199480.199515

16. Kahramanogullari, O.: Towards planning as concurrency. In: Hamza, M.H. (ed.) Artificial Intelligence and Applications, pp. 387–393. IASTED/ACTA Press, Orlando (2005)

17. Kahramanoğulları, O.: On linear logic planning and concurrency. In: Martín-Vide, C., Otto, F., Fernau, H. (eds.) LATA 2008. LNCS, vol. 5196, pp. 250–262. Springer, Heidelberg (2008)

18. Lin, F.: Applications of the situation calculus to formalizing control and strategic information: the prolog cut operator. Artif. Intell. **103**(1–2), 273–294 (1998). http://dx.doi.org/10.1016/S0004-3702(98)00054-X

19. Lloyd, J.W.: Foundations of Logic Programming. Springer-Verlag, New York (1984)

20. Nau, D., Ghallab, M., Traverso, P.: Automated Planning: Theory & Practice. Morgan Kaufmann Publishers Inc., San Francisco (2004)

21. Nguyen, T.A., Kambhampati, S.: A heuristic approach to planning with incomplete STRIPS action models. In: Chien, S., Do, M.B., Fern, A., Ruml, W. (eds.) ICAPS 2014. AAAI, Portsmouth (2014). http://www.aaai.org/ocs/index.php/ICAPS/ICAPS14/paper/view/7919

22. de Nijs, F., Klos, T.: A novel priority rule heuristic: learning from justification. In: Chien, S., Do, M.B., Fern, A., Ruml, W. (eds.) ICAPS 2014. AAAI, Portsmouth (2014). http://www.aaai.org/ocs/index.php/ICAPS/ICAPS14/paper/view/7935

23. Nilsson, N.: Principles of Artificial Intelligence. Tioga Publication Co., Paolo Alto (1980)

24. Pollock, J.L.: The logical foundations of goal-regression planning in autonomous agents. Artif. Intell. **106**(2), 267–334 (1998). http://dx.doi.org/10.1016/S0004-3702(98)00100-3

25. Pommerening, F., Röger, G., Helmert, M., Bonet, B.: Lp-based heuristics for cost-optimal planning. In: Chien, S., Do, M.B., Fern, A., Ruml, W. (eds.) ICAPS 2014. AAAI, Portsmouth (2014). http://www.aaai.org/ocs/index.php/ICAPS/ICAPS14/paper/view/7892

26. Reiter, R.: Knowledge in Action: Logical Foundations for Describing and Implementing Dynamical Systems. MIT Press, Cambridge (2001)
27. Rossi, F. (ed.) Proceedings of the 23rd International Joint Conference on Artificial Intelligence, IJCAI 2013, Beijing, China, August 3–9, 2013. IJCAI/AAAI (2013)
28. Shoham, Y.: Artificial Intelligence Techniques in Prolog. Morgan Kaufmann, New York (2014)
29. Sierra-Santibáñez, J.: Declarative formalization of strategies for action selection: applications to planning. In: Brewka, G., Moniz Pereira, L., Ojeda-Aciego, M., de Guzmán, I.P. (eds.) JELIA 2000. LNCS (LNAI), vol. 1919, p. 133. Springer, Heidelberg (2000)
30. Srivastava, S., Immerman, N., Zilberstein, S., Zhang, T.: Directed search for generalized plans using classical planners. In: Proceedings of the 21st International Conference on Automated Planning and Scheduling (ICAPS-2011). AAAI, June 2011

Web Ontology Representation and Reasoning via Fragments of Set Theory

Domenico Cantone, Cristiano Longo, Marianna Nicolosi-Asmundo$^{(\boxtimes)}$, and Daniele Francesco Santamaria

Department of Mathematics and Computer Science,
University of Catania, Catania, Italy
{cantone,longo,nicolosi}@dmi.unict.it, daniele.f.santamaria@gmail.com

Abstract. In this paper we use results from Computable Set Theory as a means to represent and reason about description logics and rule languages for the semantic web.

Specifically, we introduce the description logic $\mathcal{DL}\langle 4LQS^R\rangle(\mathbf{D})$– allowing features such as min/max cardinality constructs on the left-hand/right-hand side of inclusion axioms, role chain axioms, and datatypes–which turn out to be quite expressive if compared with $\mathcal{SROIQ}(\mathbf{D})$, the description logic underpinning the Web Ontology Language OWL. Then we show that the consistency problem for $\mathcal{DL}\langle 4LQS^R\rangle(\mathbf{D})$-knowledge bases is decidable by reducing it, through a suitable translation process, to the satisfiability problem of the stratified fragment $4LQS^R$ of set theory, involving variables of four sorts and a restricted form of quantification. We prove also that, under suitable not very restrictive constraints, the consistency problem for $\mathcal{DL}\langle 4LQS^R\rangle(\mathbf{D})$-knowledge bases is **NP**-complete. Finally, we provide a $4LQS^R$-translation of rules belonging to the Semantic Web Rule Language (SWRL).

1 Introduction

Computable Set Theory is a research field started in the late seventies with the purpose of studying the decidability of the satisfiability problem for fragments of set theory. The most efficient decision procedures designed in this area have been implemented within the reasoner Ætnanova/Referee [19] and constitute its inferential core. A wide collection of decidability results obtained up to 2001 can be found in the monographs [2,10].

Most of the decidability results and applications in computable set theory concern one-sorted multi-level syllogistics, namely collections of formulae admitting variables of one sort only, which range over the Von Neumann universe of sets. Only a few stratified syllogistics, where variables of multiple sorts are allowed, have been investigated, despite the fact that in many fields of computer

Work partially supported by the FIR project *COMPACT: Computazione affidabile su testi firmati*, code: D84C46.

B. ten Cate and A. Mileo (Eds.): RR 2015, LNCS 9209, pp. 61–76, 2015.
DOI: 10.1007/978-3-319-22002-4_6

science and mathematics one often has to deal with multi-sorted languages. For instance, in *Description Logics* one has to consider entities of different types such as *individual elements*, *concepts*, namely sets of individuals, and *roles*, namely binary relations over elements.

Recently, one-sorted multi-level fragments of set theory allowing one to express constructs related to *multi-valued maps* have been studied (see [3–5]) and applied in the realm of knowledge representation. In [7], for instance, an expressive description logic, called $\mathcal{DL}\langle \mathsf{MLSS}^{\times}_{2,m}\rangle$, has been introduced and the consistency problem for $\mathcal{DL}\langle \mathsf{MLSS}^{\times}_{2,m}\rangle$-knowledge bases has been proved **NP**-complete. $\mathcal{DL}\langle \mathsf{MLSS}^{\times}_{2,m}\rangle$ has been extended with additional description logic constructs and SWRL rules in [4], proving that the decision problem for the resulting description logic, called $\mathcal{DL}\langle \forall^{\pi}_{0,2}\rangle$, is still **NP**-complete under certain conditions. Finally, in [3] $\mathcal{DL}\langle \forall^{\pi}_{0,2}\rangle$ has been extended with some *metamodelling* features. However, none of the above-mentioned description logics provides any functionality to deal with datatypes, a simple form of concrete domains that are relevant in real-world applications.

In this paper we introduce an expressive description logic, $\mathcal{DL}\langle 4LQS^R\rangle(\mathbf{D})$ (more simply referred to as $\mathcal{DL}^4_{\mathbf{D}}$ in the rest of the paper), that can be represented in the decidable four-level stratified fragment of set theory $4LQS^R$. The logic $\mathcal{DL}^4_{\mathbf{D}}$ supports datatypes, and admits concept constructs such as full negation, union and intersection of concepts, concept domain and range, existential quantification and min cardinality on the left-hand side of inclusion axioms. It also supports role constructs such as role chains on the left hand side of inclusion axioms, union, intersection, and complement of roles, and properties on roles such as transitivity, symmetry, reflexivity, and irreflexivity.

We shall prove that the consistency problem for $\mathcal{DL}^4_{\mathbf{D}}$-knowledge bases is decidable via a reduction to the satisfiability problem for formulae of $4LQS^R$. The latter problem was proved decidable in [8]. We shall also show that the consistency problem for $\mathcal{DL}^4_{\mathbf{D}}$-knowledge bases involving only suitably constrained $\mathcal{DL}^4_{\mathbf{D}}$-formulae is **NP**-complete. Such restrictions are not very limitative: in fact, it turns out that the constrained logic allows one to represent real world ontologies such as Ontoceramic, designed for ancient ceramic cataloguing in collaboration with archaeological experts (see [9,18]).

The logic $\mathcal{DL}^4_{\mathbf{D}}$ is not an extension of $\mathcal{SROIQ}(\mathbf{D})$, the description logic upon which the W3C standard OWL 2 DL is based, as it admits existential (resp., universal) quantification only on the left-hand (resp., right-hand) side of inclusion axioms. However, $\mathcal{DL}^4_{\mathbf{D}}$ allows one to express chain axioms not supported by $\mathcal{SROIQ}(\mathbf{D})$, as they can involve roles that are not subject to any regularity restriction. Moreover, Boolean combination of roles is admitted even on the right-hand side of chain axioms. The latter fact is particularly relevant to the problem of expressing rules in OWL. We will briefly illustrate how $4LQS^R$ can be used to express SWRL rules in Sect. 3.1.

The paper is organized as follows. In Sect. 2 we review the syntax and semantics of the set-theoretic fragment $4LQS^R$ and of the logic $\mathcal{SROIQ}(\mathbf{D})$. Then, in Sect. 3, we present the description logic $\mathcal{DL}^4_{\mathbf{D}}$ and prove that the decidability of the consistency problem for $\mathcal{DL}^4_\mathbf{D}$-knowledge bases can be reduced to the satisfiability problem for $4LQS^R$-formulae. In particular, in Sect. 3.1 we show that SWRL rules can be represented within the $4LQS^R$-fragment. Finally, in Sect. 4 we draw our conclusions and give some hints to future work.

2   Preliminaries

In this section we introduce concepts and notions that will be used in the paper.

2.1   The Set-Theoretic Fragment $4LQS^R$

In order to define the fragment $4LQS^R$, it is convenient to first introduce the syntax and semantics of a more general four-level quantified language, denoted $4LQS$. Then we provide some restrictions on quantified formulae of $4LQS$ that characterize $4LQS^R$. We recall that the satisfiability problem for $4LQS^R$ has been proved decidable in [8].

$4LQS$ involves the four collections of variables $\mathcal{V}_0, \mathcal{V}_1, \mathcal{V}_2, \mathcal{V}_3$, where:

- $\mathcal{V}_0$ contains variables of sort 0, denoted by $x, y, z, ...$;
- $\mathcal{V}_1$ contains variables of sort 1, denoted by $X^1, Y^1, Z^1, ...$;
- $\mathcal{V}_2$ contains variables of sort 2, denoted by $X^2, Y^2, Z^2, ...$;
- $\mathcal{V}_3$ contains variables of sort 3, denoted by $X^3, Y^3, Z^3, ...$.

In addition to variables, $4LQS$ involves also *pair terms* of the form $\langle x, y \rangle$, for $x, y \in \mathcal{V}_0$. *$4LQS$-quantifier-free atomic formulae* are classified as:

- level 0: $x = y$, $x \in X^1$, $\langle x, y \rangle = X^2$, $\langle x, y \rangle \in X^3$, where $x, y \in \mathcal{V}_0$, $\langle x, y \rangle$ is a pair term, $X^1 \in \mathcal{V}_1$, $X^2 \in \mathcal{V}_2$, X^3 in $\mathcal{V}_3$;
- level 1: $X^1 = Y^1$, $X^1 \in X^2$, with $X^1, Y^1 \in \mathcal{V}_1$, X^2 in $\mathcal{V}_2$;
- level 2: $X^2 = Y^2$, $X^2 \in X^3$, with $X^2, Y^2 \in \mathcal{V}_2$, X^3 in $\mathcal{V}_3$.

$4LQS$ purely universal formulae are classified as:

- level 1: $(\forall z_1)...(\forall z_n)\varphi_0$, where $z_1, .., z_n \in \mathcal{V}_0$ and φ_0 is any propositional combination of quantifier-free atomic formulae of level 0;
- level 2: $(\forall Z^1_1)...(\forall Z^1_m)\varphi_1$, where $Z^1_1, .., Z^1_m \in \mathcal{V}_1$ and φ_1 is any propositional combination of quantifier-free atomic formulae of levels 0 and 1 and of purely universal formulae of level 1;
- level 3: $(\forall Z^2_1)...(\forall Z^2_p)\varphi_2$, where $Z^2_1, .., Z^2_p \in \mathcal{V}_2$ and φ_2 is any propositional combination of quantifier-free atomic formulae and of purely universal formulae of levels 1 and 2.

$4LQS$-formulae are all the propositional combinations of quantifier-free atomic formulae of levels 0, 1, 2 and of purely universal formulae of levels 1, 2, 3.

Let φ be a $4LQS$-formula. Without loss of generality, we can assume that φ contains only $\neg, \wedge, \vee$ as propositional connectives. Further, let S_φ be the syntax tree for a $4LQS$-formula $\varphi,$[1] and let ν be a node of S_φ. We say that a $4LQS$-formula ψ occurs within φ at position ν if the subtree of S_φ rooted at ν is identical to S_ψ. In this case we refer to ν as an occurrence of ψ in φ and to the path from the root of S_φ to ν as its occurrence path. An occurrence of ψ within φ is *positive* if its occurrence path deprived by its last node contains an even number of nodes labelled by a $4LQS$-formula of type $\neg\chi$. Otherwise, the occurrence is said to be *negative*.

A $4LQS$-*interpretation* is a pair $\mathcal{M} = (D, M)$ where D is any non-empty collection of objects (called domain or universe of $\mathcal{M}$) and M is an assignment over variables in $\mathcal{V}_0$, $\mathcal{V}_1$, $\mathcal{V}_2$, $\mathcal{V}_3$ such that:

  - $Mx \in D$, for each $x \in \mathcal{V}_0$; $MX^1 \in pow(D)$, for each $X^1 \in \mathcal{V}_1$;
  - $MX^2 \in pow(pow(D))$, for each $X^2 \in \mathcal{V}_2$;
  - $MX^3 \in pow(pow(pow(D)))$, for each $X^3 \in \mathcal{V}_3$

  (we recall that $pow(s)$ denotes the powerset of s).

We assume that pair terms are interpreted *à la* Kuratowski, and therefore we put $M\langle x, y\rangle =_{Def} \{\{Mx\}, \{Mx, My\}\}$. The presence of a pairing operator in the language is very useful for the set-theoretic representation of the logic $\mathcal{DL}_\mathbf{D}^4$ and of SWRL rules introduced in Sects. 3 and 3.1, respectively. Moreover, even though several pairing operators are available (see [12]), encoding ordered pairs à la Kuratowski turns out to be quite straightforward, at least for our purposes.

Next, let

  - $\mathcal{M} = (D, M)$ be a $4LQS$-interpretation,
  - $x_1, \ldots x_n \in \mathcal{V}_0$, $X_1^1, \ldots X_m^1 \in \mathcal{V}_1$, $X_1^2, \ldots X_p^2 \in \mathcal{V}_2$,
  - $u_1, \ldots u_n \in D$, $U_1^1, \ldots U_m^1 \in pow(D)$, $U_1^2, \ldots U_p^2 \in pow(pow(D))$.

By $\mathcal{M}[x_1/u_1, \ldots, x_n/u_n, X_1^1/U_1^1, \ldots X_m^1/U_m^1, X_1^2/U_1^2, \ldots X_p^2/U_p^2]$, we denote the interpretation $\mathcal{M}' = (D, M')$ such that $M'x_i = u_i$, for $i = 1, \ldots, n$, $M'X_j^1 = U_j^1$, for $j = 1, \ldots, m$, $M'X_k^2 = U_k^2$, for $k = 1, \ldots, p$, and which otherwise coincides with M on all remaining variables. Let φ be a $4LQS$-formula and let $\mathcal{M} = (D, M)$ be a $4LQS$-interpretation. The notion of satisfiability of φ by $\mathcal{M}$ (denoted by $\mathcal{M} \models \varphi$) is defined inductively over the structure of φ. Quantifier-free atomic formulae are evaluated in a standard way according to the usual meaning of the predicates '$\in$' and '$=$', and purely universal formulae are evaluated as follows:

  - $\mathcal{M} \models (\forall z_1)\ldots(\forall z_n)\varphi_0$ iff $\mathcal{M}[z_1/u_1, \ldots, z_n/u_n] \models \varphi_0$, for all $u_1, \ldots u_n \in D$;
  - $\mathcal{M} \models (\forall Z_1^1)\ldots(\forall Z_m^1)\varphi_1$ iff $\mathcal{M}[Z_1^1/U_1^1, \ldots, Z_n^1/U_n^1] \models \varphi_1$, for all $U_1^1, \ldots U_m^1 \in pow(D)$;
  - $\mathcal{M} \models (\forall Z_1^2)\ldots(\forall Z_m^2)\varphi_2$ iff $\mathcal{M}[Z_1^2/U_1^2, \ldots, Z_n^2/U_n^2] \models \varphi_2$, for all $U_1^2, \ldots U_m^2 \in pow(pow(D))$.

Finally, compound formulae are interpreted according to the standard rules of propositional logic. If $\mathcal{M} \models \varphi$, then $\mathcal{M}$ is said to be a $4LQS$-model for φ. A

---

[1] The notion of syntax tree for $4LQS$-formulae is similar to the notion of syntax tree for formulae of first-order logic. A precise definition of the latter can be found in [11].

4LQS-formula is said to be *satisfiable* if it has a 4LQS-model. A 4LQS-formula is *valid* if it is satisfied by all 4LQS-interpretations.

Next we present the fragment $4LQS^R$ of 4LQS of our interest, namely the collection of the formulae ψ of 4LQS fulfilling the restrictions:

1. for every purely universal formula $(\forall Z_1^1), ..., (\forall Z_m^1)\varphi_1$ of level 2 occurring in ψ and every purely universal formula $(\forall z_1), ..., (\forall z_n)\varphi_0$ of level 1 occurring negatively in φ_1, the condition

$$\neg\varphi_0 \rightarrow \bigwedge_{i=1}^{n} \bigwedge_{j=1}^{m} z_i \in Z_j^1$$

   is a valid 4LQS-formula (in this case we say that $(\forall z_1), ..., (\forall z_n)\varphi_0$ is *linked to the variables* $Z_1^1, ..., Z_m^1$);
2. for every purely universal formula $(\forall Z_1^2), ..., (\forall Z_p^2)\varphi_2$ of level 3 in ψ:
   - every purely universal formula of level 1 occurring negatively in φ_2 and not occurring in a purely universal formula of level 2 is only allowed to be of the form

$$(\forall z_1), ..., (\forall z_n)\neg(\bigwedge_{i=1}^{n} \bigwedge_{j=1}^{n} \langle z_i, z_j \rangle = Y_{ij}^2),$$

   with $Y_{ij}^2 \in \mathcal{V}^2$, for $i, j = 1, ..., n$;
   - purely universal formulae $(\forall Z_1^1), ..., (\forall Z_m^1)\varphi_1$ of level 2 may occur only positively in φ_2.

Restriction 1 has been introduced for technical reasons concerning the decidability of the satisfiability problem for the fragment. In fact it guarantees that satisfiability is preserved in a suitable finite submodel of ψ. Restriction 2 allows one to express binary relations and several operations on them while keeping simple, at the same time, the decision procedure (for space reasons details are not included here but can be found in [8]).

We observe that the semantics of $4LQS^R$ plainly coincides with that of 4LQS.

In the $4LQS^R$-fragment one can express several set-theoretic constructs such as a restricted variant of the set former, which in turn allows one to express other significant set operators such as binary union, intersection, set difference, the singleton operator, the powerset operator, etc. Within the fragment $4LQS^R$, it is also possible to define binary relations over elements of a domain together with conditions on them (i.e., reflexivity, transitivity, weak connectedness, irreflexivity, intransitivity) which characterize accessibility relations of well-known modal logics. In particular, the normal modal logic K45 can be translated in the $4LQS^R$-fragment. The interested reader is referred to [8] for details. The modal logic S5 can be represented in $4LQS^R$ much in a similar way.

2.2 Description Logics

Description Logics (DL) are a family of formalisms widely used in the field of Knowledge Representation to model application domains and to reason on them [1]. DL knowledge bases describe models that are based on individual elements (or, more simply, individuals), classes whose elements are individual names, and binary relationships between individuals. One of the leading application domains for DL is the semantic web. In fact, the most recently developed semantic web language, namely OWL 2, is based on a very expressive description logic with datatypes $\mathbf{D}$, called $\mathcal{SROIQ}(\mathbf{D})$. Extensions of DL with datatypes have been studied and analyzed in [14,17].

The logic $\mathcal{SROIQ}(\mathbf{D})$ is briefly introduced in the next section (the interested reader is referred to [13] for details).

2.2.1   The Description Logic $\mathcal{SROIQ}(\mathbf{D})$

Let $\mathbf{D} = (N_D, N_C, N_F, \cdot^{\mathbf{D}})$ be a *datatype map* in the sense of [17], where N_D is a finite set of datatypes, N_C is a function assigning a set of constants $N_C(d)$ to each datatype $d \in N_D$, N_F is a function assigning a set of facets $N_F(d)$ to each $d \in N_D$, and $\cdot^{\mathbf{D}}$ is a function assigning a datatype interpretation $d^{\mathbf{D}}$ to each datatype $d \in N_D$, a facet interpretation $f^{\mathbf{D}} \subseteq d^{\mathbf{D}}$ to each facet $f \in N_F(d)$, and a data value $e_d^{\mathbf{D}} \in d^{\mathbf{D}}$ to every constant $e_d \in N_C(d)$. We shall assume that the interpretations of the datatypes in N_D are nonempty pairwise disjoint sets.

A *facet expression* for a datatype $d \in N_D$ is a formula ψ_d constructed from the elements of $N_F(d) \cup \{\top_d, \perp_d\}$ by applying a finite number of times the connectives $\neg$, $\wedge$, and $\vee$. The function $\cdot^{\mathbf{D}}$ is extended to facet expressions for $d \in N_D$ by putting $\top_d^{\mathbf{D}} = d^{\mathbf{D}}$, $\perp_d^{\mathbf{D}} = \emptyset$, $(\neg f)^{\mathbf{D}} = d^{\mathbf{D}} \setminus f^{\mathbf{D}}$, $(f_1 \wedge f_2)^{\mathbf{D}} = f_1^{\mathbf{D}} \cap f_2^{\mathbf{D}}$, and $(f_1 \vee f_2)^{\mathbf{D}} = f_1^{\mathbf{D}} \cup f_2^{\mathbf{D}}$, for $f, f_1, f_2 \in N_F(d)$.

A *data range dr* for $\mathbf{D}$ is either a datatype $d \in N_D$, or a finite enumeration of datatype constants $\{e_{d_1}, \dots, e_{d_n}\}$, with $e_{d_i} \in N_C(d_i)$ and $d_i \in N_D$, or a facet expression ψ_d, for $d \in N_D$, or their negation.

Let $\mathbf{R_A}$, $\mathbf{R_D}$, $\mathbf{C}$, $\mathbf{I}$ be denumerable pairwise disjoint sets of abstract role names, concrete role names, concept names, and individual names, respectively. The set of abstract roles is defined as $\mathbf{R_A} \cup \{R^- \mid R \in \mathbf{R_A}\} \cup U$, where U is the universal role and R^- is the inverse role of R. A role inclusion axiom (RIA) is an expression of the form $w \sqsubseteq R$, where w is a finite string of roles not including U and R is an abstract role name distinct from U. An abstract role hierarchy R_a^H is a finite collection of RIAs. A concrete role hierarchy $\mathsf{R_D}^H$ is a finite collection of concrete role inclusion axioms $T_i \sqsubseteq T_j$, where $T_i, T_j \in \mathbf{R_D}$. A role assertion is an expression of one of the types: $\mathsf{Ref}(R)$, $\mathsf{Irref}(R)$, $\mathsf{Sym}(R)$, $\mathsf{Asym}(R)$, $\mathsf{Tra}(R)$, and $\mathsf{Dis}(R, S)$, where $R, S \in \mathbf{R_A} \cup \{R^- \mid R \in \mathbf{R_A}\}$.

Given an abstract role hierarchy R_a^H and a set of role assertions R^A without transitivity or symmetry assertions ($\mathsf{Sym}(R)$ can be represented by a RIA of type $R^- \sqsubseteq R$ and $\mathsf{Tra}(R)$ by $RR \sqsubseteq R$), the set of roles that are *simple* in $\mathsf{R}_a^H \cup \mathsf{R}^A$ is inductively defined as follows: (a) a role name is simple if it does not occur on the right hand side of a RIA in R_a^H, (b) an inverse role R^- is simple if R is, and (c) if R occurs on the right hand of a RIA in R_a^H, then R is simple if, for each $w \sqsubseteq R \in \mathsf{R}_a^H$, $w = S$, for a simple role S.

A set of role assertions R^A is called simple if all roles R, S appearing in role assertions of the form $\mathsf{Irref}(R)$, $\mathsf{Asym}(R)$, or $\mathsf{Dis}(R, S)$ are simple in R^A.

An $\mathcal{SROIQ}(\mathbf{D})$-*RBox* is a set $\mathsf{R} = \mathsf{R}_a^H \cup \mathsf{R}_\mathbf{D}^H \cup \mathsf{R}^A$ such that R_a^H is a regular abstract role hierarchy, $\mathsf{R}_\mathbf{D}^H$ is a concrete role hierarchy, and R^A is a finite simple set of role assertions. A formal definition of regular abstract role hierarchy can be found in [13].

Before introducing the formal definitions of *TBox* and of *ABox*, we define the set of $\mathcal{SROIQ}(\mathbf{D})$-concepts as the smallest set such that:

- every concept name and the constants $\top$, $\bot$ are concepts;
- if C, D are concepts, R is an abstract role (possibly inverse), S is a simple role (possibly inverse), T is a concrete role, dr is a data range for $\mathbf{D}$, a is an individual, and n is a non-negative integer, then $C \sqcap D$, $C \sqcup D$, $\neg C$, $\{a\}$, $\forall R.C$, $\exists R.C$, $\exists S.Self$, $\forall T.dr$, $\exists T.dr$, $\geq nS.C$, and $\leq nS.C$ are also concepts.

A general concept inclusion axiom (GCI) is an expression $C \sqsubseteq D$, where C, D are $\mathcal{SROIQ}(\mathbf{D})$-concepts. An $\mathcal{SROIQ}(\mathbf{D})$-*TBox* $\mathcal{T}$ is a finite set of CGIs.

Any expression of one of the following forms: $a : C$, $(a, b) : R$, $(a, e_d) : T$, $(a, b) : \neg R$, $(a, e_d) : \neg T$, $a = b$, $a \neq b$, where a, b are individuals, e_d is a constant in $N_C(d)$, R is a (possibly) inverse abstract role, P is a concrete role, and C is a concept, is called an *individual assertion*. An $\mathcal{SROIQ}(\mathbf{D})$-*ABox* $\mathcal{A}$ is a finite set of individual assertions.

An $\mathcal{SROIQ}(\mathbf{D})$-knowledge base is a triple $\mathcal{K} = (\mathcal{R}, \mathcal{T}, \mathcal{A})$ such that $\mathcal{R}$ is an $\mathcal{SROIQ}(\mathbf{D})$-*RBox*, $\mathcal{T}$ an $\mathcal{SROIQ}(\mathbf{D})$-*TBox*, and $\mathcal{A}$ an $\mathcal{SROIQ}(\mathbf{D})$-*ABox*. The semantics of $\mathcal{SROIQ}(\mathbf{D})$ is given by means of an interpretation $\mathbf{I} = (\Delta^{\mathbf{I}}, \Delta_{\mathbf{D}}, \cdot^{\mathbf{I}})$, where $\Delta^{\mathbf{I}}$ and $\Delta_{\mathbf{D}}$ are non-empty disjoint domains such that $d^{\mathbf{D}} \subseteq \Delta_{\mathbf{D}}$, for every $d \in N_D$, and $\cdot^{\mathbf{I}}$ is an interpretation function. For space reasons, the definition of the interpretation of concepts and roles, axioms and assertions is not reported here. However, it can be found in [6, Table 1].

Let $\mathcal{R}, \mathcal{T}$, and $\mathcal{A}$ be as above. An interpretation $\mathbf{I} = (\Delta^{\mathbf{I}}, \Delta_{\mathbf{D}}, \cdot^{\mathbf{I}})$ is a $\mathbf{D}$-model of $\mathcal{R}$ (resp., $\mathcal{T}$), and we write $\mathbf{I} \models_{\mathbf{D}} \mathcal{R}$ (resp., $\mathbf{I} \models_{\mathbf{D}} \mathcal{T}$), if $\mathbf{I}$ satisfies each axiom in $\mathcal{R}$ (resp., $\mathcal{T}$) according to the semantic rules in [6, Table 1]. Analogously, $\mathbf{I} = (\Delta^{\mathbf{I}}, \Delta_{\mathbf{D}}, \cdot^{\mathbf{I}})$ is a $\mathbf{D}$-model of $\mathcal{A}$, and we write $\mathbf{I} \models_{\mathbf{D}} \mathcal{A}$, if $\mathbf{I}$ satisfies each assertion in $\mathcal{A}$, according to the semantic rules in [6, Table 1].

An $\mathcal{SROIQ}(\mathbf{D})$-knowledge base $\mathcal{K} = (\mathcal{A}, \mathcal{T}, \mathcal{R})$ is consistent if there is an interpretation $\mathbf{I} = (\Delta^{\mathbf{I}}, \Delta_{\mathbf{D}}, \cdot^{\mathbf{I}})$ that is a $\mathbf{D}$-model of $\mathcal{A}, \mathcal{T}$, and $\mathcal{R}$.

Decidability of the consistency problem for $\mathcal{SROIQ}(\mathbf{D})$-knowledge bases was proved in [13] by means of a tableau-based decision procedure and its computational complexity was shown to be **N2EXPTime**-complete in [15].

3   The Logic $\mathcal{DL}\langle 4LQS^R \rangle(\mathbf{D})$

In this section we introduce the description logic $\mathcal{DL}\langle 4LQS^R \rangle(\mathbf{D})$ (shortly referred to as $\mathcal{DL}_{\mathbf{D}}^4$) and prove that the consistency problem for $\mathcal{DL}_{\mathbf{D}}^4$-knowledge bases is decidable by reducing it to the satisfiability problem for $4LQS^R$-formulae. Then we show that under certain restrictions the consistency problem for $\mathcal{DL}_{\mathbf{D}}^4$-knowledge bases is **NP**-complete. Finally we briefly illustrate how SWRL-rules can be translated into the language of $4LQS^R$.

Let $\mathbf{D}, \mathbf{R_A}, \mathbf{R_D}, \mathbf{I}, \mathbf{C}$ be as in Sect. 2.2.1.
(a) $\mathcal{DL}_{\mathbf{D}}^4$-datatype, (b) $\mathcal{DL}_{\mathbf{D}}^4$-concept, (c) $\mathcal{DL}_{\mathbf{D}}^4$-abstract role, and (d) $\mathcal{DL}_{\mathbf{D}}^4$-concrete role terms are constructed according to the following syntax rules:

(a) $t_1, t_2 \longrightarrow dr \mid \neg t_1 \mid t_1 \sqcap t_2 \mid t_1 \sqcup t_2 \mid \{e_d\}$,
(b) $C_1, C_2 \longrightarrow A \mid \top \mid \bot \mid \neg C_1 \mid C_1 \sqcup C_2 \mid C_1 \sqcap C_2 \mid \{a\} \mid \exists R.Self \mid \exists R.\{a\} \mid \exists P.\{e_d\}$,
(c) $R_1, R_2 \longrightarrow S \mid U \mid R_1^- \mid \neg R_1 \mid R_1 \sqcup R_2 \mid R_1 \sqcap R_2 \mid R_{C_1 \mid} \mid R_{\mid C_1} \mid R_{C_1 \mid C_2} \mid id(C)$,
(d) $P \longrightarrow T \mid \neg P \mid P_{C_1 \mid} \mid P_{\mid t_1} \mid P_{C_1 \mid t_1}$,

where dr is a data range for $\mathbf{D}$, t_1, t_2 are datatype terms, e_d is a constant in $N_C(d)$, a is an individual name, A is a concept name, C_1, C_2 are $\mathcal{DL}_{\mathbf{D}}^4$-concept terms, S is an abstract role name, R, R_1, R_2 are $\mathcal{DL}_{\mathbf{D}}^4$-abstract role terms, T a concrete role name, and P a $\mathcal{DL}_{\mathbf{D}}^4$-concrete role term.

A $\mathcal{DL}_{\mathbf{D}}^4$-knowledge base is a triple $\mathcal{K} = (\mathcal{R}, \mathcal{T}, \mathcal{A})$ such that $\mathcal{R}$ is a $\mathcal{DL}_{\mathbf{D}}^4$-RBox, $\mathcal{T}$ is a $\mathcal{DL}_{\mathbf{D}}^4$-TBox, and $\mathcal{A}$ a $\mathcal{DL}_{\mathbf{D}}^4$-ABox. A $\mathcal{DL}_{\mathbf{D}}^4$-RBox is a collection of statements of the following forms: $R_1 \equiv R_2$, $R_1 \sqsubseteq R_2$, $R_1 \ldots R_n \sqsubseteq R_{n+1}$, $\mathsf{Sym}(R_1)$, $\mathsf{Asym}(R_1)$, $\mathsf{Ref}(R_1)$, $\mathsf{Irref}(R_1)$, $\mathsf{Dis}(R_1, R_2)$, $\mathsf{Tra}(R_1)$, $\mathsf{Fun}(R_1)$, $P_1 \equiv P_2$, $P_1 \sqsubseteq P_2$, $\mathsf{Fun}(P_1)$, where R_1, R_2 are $\mathcal{DL}_{\mathbf{D}}^4$-abstract role terms and P_1, P_2 are $\mathcal{DL}_{\mathbf{D}}^4$-concrete role terms. A $\mathcal{DL}_{\mathbf{D}}^4$-TBox is a set of statements of the types:

- $C_1 \equiv C_2$, $C_1 \sqsubseteq C_2$, $C_1 \sqsubseteq \forall R_1.C_2$, $\exists R_1.C_1 \sqsubseteq C_2$, $\geq_n R_1.C_1 \sqsubseteq C_2$,
  $C_1 \sqsubseteq \leq_n R_1.C_2$,
- $t_1 \equiv t_2$, $t_1 \sqsubseteq t_2$, $C_1 \sqsubseteq \forall P_1.t_1$, $\exists P_1.t_1 \sqsubseteq C_1$, $\geq_n P_1.t_1 \sqsubseteq C_1$, $C_1 \sqsubseteq \leq_n P_1.t_1$,

where C_1, C_2 are $\mathcal{DL}_{\mathbf{D}}^4$-concept terms, t_1, t_2 datatype terms, R_1 a $\mathcal{DL}_{\mathbf{D}}^4$-abstract role term, P_1 a $\mathcal{DL}_{\mathbf{D}}^4$-concrete role term.

A $\mathcal{DL}_{\mathbf{D}}^4$-ABox is a set of assertions of the forms: $a : C_1$, $(a, b) : R_1$, $(a, b) : \neg R_1$, $a = b$, $a \neq b$, $e_d : t_1$, $(a, e_d) : P_1$, $(a, e_d) : \neg P_1$, with C_1 a $\mathcal{DL}_{\mathbf{D}}^4$-concept term, d a datatype, t_1 a datatype term, R_1 a $\mathcal{DL}_{\mathbf{D}}^4$-abstract role term, P_1 a $\mathcal{DL}_{\mathbf{D}}^4$-concrete role term, a, b individual names, and e_d a constant in $N_C(d)$.

The semantics of $\mathcal{DL}_{\mathbf{D}}^4$ is similar to that of $\mathcal{SROIQ}(\mathbf{D})$. The interpretation of terms, axioms, and assertions of $\mathcal{DL}_{\mathbf{D}}^4$ shared with $\mathcal{SROIQ}(\mathbf{D})$ is illustrated in [6, Table 1], while the semantics of terms and statements specific to $\mathcal{DL}_{\mathbf{D}}^4$ is described in [6, Table 2]. The notions of $\mathbf{D}$-model of a $\mathcal{DL}_{\mathbf{D}}^4$-RBox, $\mathcal{DL}_{\mathbf{D}}^4$-TBox, $\mathcal{DL}_{\mathbf{D}}^4$-ABox, and the notion of consistency of a $\mathcal{DL}_{\mathbf{D}}^4$-knowledge base are similar to the ones described in [6, Table 1] for $\mathcal{SROIQ}(\mathbf{D})$.

In the following theorem we prove the decidability of the consistency problem for $\mathcal{DL}_{\mathbf{D}}^4$-knowledge bases.

Theorem 1. *Let $\mathcal{K}$ be a $\mathcal{DL}_{\mathbf{D}}^4$-knowledge base. Then, one can construct a $4LQS^R$-formula $\varphi_{\mathcal{K}}$ s. t. $\varphi_{\mathcal{K}}$ is satisfiable if and only if $\mathcal{K}$ is consistent.*

Proof. As a preliminary step, observe that the statements of the $\mathcal{DL}_{\mathbf{D}}^4$-knowledge base $\mathcal{K}$ that need to be considered are those of the following types:

- $C_1 \equiv \top$, $C_1 \equiv \neg C_2$, $C_1 \equiv C_2 \sqcup C_3$, $C_1 \equiv \{a\}$, $C_1 \sqsubseteq \forall R_1.C_2$, $\exists R_1.C_1 \sqsubseteq C_2$,
  $\geq_n R_1.C_1 \sqsubseteq C_2$, $C_1 \sqsubseteq \leq_n R_1.C_2$, $C_1 \sqsubseteq \forall P_1.t_1$, $\exists P_1.t_1 \sqsubseteq C_1$, $\geq_n P_1.t_1 \sqsubseteq C_1$,
  $C_1 \sqsubseteq \leq_n P_1.t_1$,

- $R_1 \equiv U$, $R_1 \equiv \neg R_2$, $R_1 \equiv R_2 \sqcup R_3$, $R_1 \equiv R_2^-$, $R_1 \equiv id(C_1)$, $R_1 \equiv R_{2_{C_1|}}$, $R_1 \ldots R_n \sqsubseteq R_{n+1}$, $\mathsf{Ref}(R_1)$, $\mathsf{Irref}(R_1)$, $\mathsf{Dis}(R_1, R_2)$, $\mathsf{Fun}(R_1)$,
- $P_1 \equiv P_2$, $P_1 \equiv \neg P_2$, $P_1 \sqsubseteq P_2$, $\mathsf{Fun}(P_1)$, $P_1 \equiv P_{2_{C_1|}}$, $P_1 \equiv P_{2_{C_1|t_1}}$, $P_1 \equiv P_{2_{|t_1}}$,
- $t_1 \equiv t_2$, $t_1 \equiv \neg t_2$, $t_1 \equiv t_2 \sqcup t_3$, $t_1 \equiv \{e_d\}$,
- $a : C_1$, $(a, b) : R_1$, $(a, b) : \neg R_1$, $a = b$, $a \neq b$, $e_d : t_1$, $(a, e_d) : P_1$, $(a, e_d) : \neg P_1$.

In order to define the $4LQS^R$-formula $\varphi_{\mathcal{K}}$, we shall make use of a mapping τ from the $\mathcal{DL}_{\mathbf{D}}^4$-statements (and their conjunctions) listed above into $4LQS^R$-formulae. To prepare for the definition of τ, we map injectively individuals a and constants $e_d \in N_C(d)$ into level 0 variables x_a and x_{e_d}, the constant concepts $\top$ and $\bot$, datatype terms t, and concept terms C into level 1 variables $X_\top^1$, $X_\bot^1$, X_t^1, X_C^1, respectively, and the universal relation on individuals U, abstract role terms R, and concrete role terms P into level 3 variables X_U^3, X_R^3, and X_P^3, respectively.[2]

Then the mapping τ is defined as follows:

$\tau(C_1 \equiv \top) =_{\mathrm{Def}} (\forall z)(z \in X_{C_1}^1 \leftrightarrow z \in X_\top^1)$,

$\tau(C_1 \equiv \neg C_2) =_{\mathrm{Def}} (\forall z)(z \in X_{C_1} \leftrightarrow \neg(z \in X_{C_2}^1))$,

$\tau(C_1 \equiv C_2 \sqcup C_3) =_{\mathrm{Def}} (\forall z)(z \in X_{C_1}^1 \leftrightarrow (z \in X_{C_2}^1 \vee z \in X_{C_3}^1))$,

$\tau(C_1 \equiv \{a\}) =_{\mathrm{Def}} (\forall z)(z \in X_{C_1}^1 \leftrightarrow z = x_a)$,

$\tau(C_1 \sqsubseteq \forall R_1.C_2) =_{\mathrm{Def}} (\forall z_1)(\forall z_2)(z_1 \in X_{C_1}^1 \rightarrow (\langle z_1, z_2 \rangle \in X_{R_1}^3 \rightarrow z_2 \in X_{C_2}^1))$,

$\tau(\exists R_1.C_1 \sqsubseteq C_2) =_{\mathrm{Def}} (\forall z_1)(\forall z_2)((\langle z_1, z_2 \rangle \in X_{R_1}^3 \wedge z_2 \in X_{C_1}^1) \rightarrow z_1 \in X_{C_2}^1)$,

$\tau(C_1 \equiv \exists R_1.\{a\}) =_{\mathrm{Def}} (\forall z)(z \in X_{C_1}^1 \leftrightarrow \langle z, x_a \rangle \in X_{R_1}^3)$,

$\tau(C_1 \sqsubseteq_{\leq n} R_1.C_2) =_{\mathrm{Def}} (\forall z)(\forall z_1) \ldots (\forall z_{n+1})(z \in X_{C_1}^1 \rightarrow$
$$(\bigwedge_{i=1}^{n+1} (z_i \in X_{C_2} \wedge \langle z, z_i \rangle \in X_{R_1}^3) \rightarrow \bigvee_{i<j} z_i = z_j)),$$

$\tau(\geq_n R_1.C_1 \sqsubseteq C_2) =_{\mathrm{Def}} (\forall z)(\forall z_1) \ldots (\forall z_n)(\bigwedge_{i=1}^n ((z_i \in X_{C_1}^1 \wedge \langle z, z_i \rangle \in X_{R_1}^3) \rightarrow$
$$\bigwedge_{i<j} z_i \neq z_j) \rightarrow z \in X_{C_2}^1),$$

$\tau(C_1 \sqsubseteq \forall P_1.t_1) =_{\mathrm{Def}} (\forall z_1)(\forall z_2)(z_1 \in X_{C_1}^1 \rightarrow (\langle z_1, z_2 \rangle \in X_{P_1}^3 \rightarrow z_2 \in X_{t_1}^1))$,

$\tau(\exists P_1.t_1 \sqsubseteq C_1) =_{\mathrm{Def}} (\forall z_1)(\forall z_2)((\langle z_1, z_2 \rangle \in X_{P_1}^3 \wedge z_2 \in X_{t_1}^1) \rightarrow z_1 \in X_{C_1}^1)$,

$\tau(C_1 \equiv \exists P_1.\{e_d\}) =_{\mathrm{Def}} (\forall z)(z \in X_{C_1}^1 \leftrightarrow \langle z, x_{e_d} \rangle \in X_{P_1}^3)$,

$\tau(C_1 \sqsubseteq_{\leq n} P_1.t_1) =_{\mathrm{Def}} (\forall z)(\forall z_1) \ldots (\forall z_{n+1})(z \in X_{C_1}^1 \rightarrow$
$$(\bigwedge_{i=1}^{n+1} (z_i \in X_{t_1} \wedge \langle z, z_i \rangle \in X_{P_1}^3) \rightarrow \bigvee_{i<j} z_i = z_j)),$$

$\tau(\geq_n P_1.t_1 \sqsubseteq C_1) =_{\mathrm{Def}}$
$$(\forall z)(\forall z_1) \ldots (\forall z_n)(\bigwedge_{i=1}^n ((z_i \in X_{t_1}^1 \wedge \langle z, z_i \rangle \in X_{P_1}^3) \rightarrow \bigwedge_{i<j} z_i \neq z_j) \rightarrow z \in X_{C_1}^1),$$

$\tau(R_1 \equiv U) =_{\mathrm{Def}} (\forall Z^2)(Z^2 \in X_{R_1}^3 \leftrightarrow Z^2 \in X_U^3)$,

$\tau(R_1 \equiv \neg R_2) =_{\mathrm{Def}} (\forall z_1)(\forall z_2)((\langle z_1, z_2 \rangle \in X_{R_1}^3 \leftrightarrow \neg(\langle z_1, z_2 \rangle \in X_{R_2}^3))$,

$\tau(R_1 \equiv R_2 \sqcup R_3) =_{\mathrm{Def}} (\forall Z^2)(Z^2 \in X_{R_1}^3 \leftrightarrow (Z^2 \in X_{R_2}^3 \vee Z^2 \in X_{R_3}^3))$,

$\tau(R_1 \equiv R_2^-) =_{\mathrm{Def}} (\forall z_1)(\forall z_2)((\langle z_1, z_2 \rangle \in X_{R_1}^3 \leftrightarrow \langle z_2, z_1 \rangle \in X_{R_2}^3))$,

---

[2] The use of level 3 variables to model abstract and concrete role terms is motivated by the fact that their elements, that is ordered pairs $\langle x, y \rangle$, are encoded in Kuratowski's style as $\{\{x\}, \{x, y\}\}$, namely as collections of sets of objects. Variables of level 2 are used in the formulae ψ_8 and ψ_9 of the construction to model the fact that level 3 variables representing role terms are binary relations.

$\tau(R_1 \equiv id(C_1)) =_{\text{Def}} (\forall z_1)(\forall z_2)(\langle z_1, z_2 \rangle \in X^3_{R_1} \leftrightarrow (z_1 \in X^1_{C_1} \wedge z_2 \in X^1_{C_1} \wedge z_1 = z_2)),$

$\tau(R_1 \equiv R_{2_{C_1|}}) =_{\text{Def}} (\forall z_1)(\forall z_2)(\langle z_1, z_2 \rangle \in X^3_{R_1} \leftrightarrow (\langle z_1, z_2 \rangle \in X^3_{R_2} \wedge z_1 \in X^1_{C_1})),$

$\tau(R_1 \ldots R_n \sqsubseteq R_{n+1}) =_{\text{Def}} (\forall z)(\forall z_1) \ldots (\forall z_n)$
$$(((\langle z, z_1 \rangle \in X^3_{R_1} \wedge \ldots \wedge \langle z_{n-1}, z_n \rangle \in X^3_{R_n}) \rightarrow \langle z, z_n \rangle \in X^3_{R_{n+1}}),$$

$\tau(\mathsf{Ref}(R_1)) =_{\text{Def}} (\forall z)(\langle z, z \rangle \in X^3_{R_1}),$

$\tau(\mathsf{Irref}(R_1)) =_{\text{Def}} (\forall z)(\neg(\langle z, z \rangle \in X^3_{R_1})),$

$\tau(\mathsf{Fun}(R_1)) =_{\text{Def}} (\forall z_1)(\forall z_2)(\forall z_3)((\langle z_1, z_2 \rangle \in X^3_{R_1} \wedge \langle z_1, z_3 \rangle \in X^3_{R_1}) \rightarrow z_2 = z_3),$

$\tau(P_1 \equiv P_2) =_{\text{Def}} (\forall Z^2)(Z^2 \in X^3_{P_1} \leftrightarrow Z^2 \in X^3_{P_2}),$

$\tau(P_1 \equiv \neg P_2) =_{\text{Def}} (\forall Z^2)(Z^2 \in X^3_{P_1} \leftrightarrow \neg(Z^2 \in X^3_{P_2})),$

$\tau(P_1 \sqsubseteq P_2) =_{\text{Def}} (\forall Z^2)(Z^2 \in X^3_{P_1} \rightarrow Z^2 \in X^3_{P_2}),$

$\tau(\mathsf{Fun}(P_1)) =_{\text{Def}} (\forall z_1)(\forall z_2)(\forall z_3)((\langle z_1, z_2 \rangle \in X^3_{P_1} \wedge \langle z_1, z_3 \rangle \in X^3_{P_1}) \rightarrow z_2 = z_3),$

$\tau(P_1 \equiv P_{2_{C_1|}}) =_{\text{Def}} (\forall z_1)(\forall z_2)(\langle z_1, z_2 \rangle \in X^3_{P_1} \leftrightarrow (\langle z_1, z_2 \rangle \in X^3_{P_2} \wedge z_1 \in X^1_{C_1})),$

$\tau(P_1 \equiv P_{2_{|t_1}}) =_{\text{Def}} (\forall z_1)(\forall z_2)(\langle z_1, z_2 \rangle \in X^3_{P_1} \leftrightarrow (\langle z_1, z_2 \rangle \in X^3_{P_2} \wedge z_2 \in X^1_{t_1})),$

$\tau(P_1 \equiv P_{2_{C_1|t_1}}) =_{\text{Def}} (\forall z_1)(\forall z_2)(\langle z_1, z_2 \rangle \in X^3_{P_1} \leftrightarrow$
$$(\langle z_1, z_2 \rangle \in X^3_{P_2} \wedge z_1 \in X^1_{C_1} \wedge z_2 \in X^1_{t_1})),$$

$\tau(t_1 \equiv t_2) =_{\text{Def}} (\forall z)(z \in X^1_{t_1} \leftrightarrow z \in X^1_{t_2}),$

$\tau(t_1 \equiv \neg t_2) =_{\text{Def}} (\forall z)(z \in X^1_{t_1} \leftrightarrow \neg(z \in X^1_{t_2})),$

$\tau(t_1 \equiv t_2 \sqcup t_3) =_{\text{Def}} (\forall z)(z \in X^1_{t_1} \leftrightarrow (z \in X^1_{t_2} \vee z \in X^1_{t_3})),$

$\tau(t_1 \equiv t_2 \sqcap t_3) =_{\text{Def}} (\forall z)(z \in X^1_{t_1} \leftrightarrow (z \in X^1_{t_2} \wedge z \in X^1_{t_3})),$

$\tau(t_1 \equiv \{e_d\}) =_{\text{Def}} (\forall z)(z \in X^1_{t_1} \leftrightarrow z = x_{e_d}),$

$\tau(a : C_1) =_{\text{Def}} x_a \in X^1_{C_1}, \tau((a, b) : R_1) =_{\text{Def}} \langle x_a, x_b \rangle \in X^3_{R_1},$

$\tau((a, b) : \neg R_1) =_{\text{Def}} \neg(\langle x_a, x_b \rangle \in X^3_{R_1}),$

$\tau(a = b) =_{\text{Def}} x_a = x_b, \tau(a \neq b) =_{\text{Def}} \neg(x_a = x_b),$

$\tau(e_d : t_1) =_{\text{Def}} x_{e_d} \in X^1_{t_1},$

$\tau((a, e_d) : P_1) =_{\text{Def}} \langle x_a, x_{e_d} \rangle \in X^3_{P_1}, \tau((a, e_d) : \neg P_1) =_{\text{Def}} \neg(\langle x_a, x_{e_d} \rangle \in X^3_{P_1}),$

$\tau(\alpha \wedge \beta) =_{\text{Def}} \tau(\alpha) \wedge \tau(\beta).$

Let $\mathcal{K}$ be our $\mathcal{DL}^4_{\mathbf{D}}$-knowledge base, and let $\mathsf{cpt}_{\mathcal{K}}$, $\mathsf{arl}_{\mathcal{K}}$, $\mathsf{crl}_{\mathcal{K}}$, and $\mathsf{ind}_{\mathcal{K}}$ be, respectively, the sets of concept, of abstract role, of concrete role, and of individual names in $\mathcal{K}$. Moreover, let $N^{\mathcal{K}}_D \subseteq N_D$ be the set of datatypes in $\mathcal{K}$, $N^{\mathcal{K}}_F$ a restriction of N_F assigning to every $d \in N^{\mathcal{K}}_{\mathbf{D}}$ the set $N^{\mathcal{K}}_F(d)$ of facets in $N_F(d)$ and in $\mathcal{K}$. Analogously, let $N^{\mathcal{K}}_C$ be a restriction of the function N_C associating to every $d \in N^{\mathcal{K}}_{\mathbf{D}}$ the set $N^{\mathcal{K}}_C(d)$ of constants contained in $N_C(d)$ and in $\mathcal{K}$. Finally, for every datatype $d \in N^{\mathcal{K}}_{\mathbf{D}}$, let $\mathsf{bf}^{\mathbf{D}}_{\mathcal{K}}(d)$ be the set of facet expressions for d occurring in $\mathcal{K}$ and not in $N_F(d) \cup \{\top^d, \bot_d\}$. We define the $4LQS^R$-formula $\varphi_{\mathcal{K}}$ expressing the consistency of $\mathcal{K}$ as follows:

$$\varphi_{\mathcal{K}} =_{\text{Def}} \bigwedge_{i=1}^{12} \psi_i \wedge \bigwedge_{H \in \mathcal{K}} \tau(H),$$

where

- $\psi_1 =_{\text{Def}} (\forall z)(z \in X^1_{\mathbf{I}} \leftrightarrow \neg(z \in X^1_{\mathbf{D}})) \wedge (\forall z)(z \in X^1_{\mathbf{I}} \vee z \in X^1_{\mathbf{D}}) \wedge$
$$\neg(\forall z)\neg(z \in X^1_{\mathbf{I}}) \wedge \neg(\forall z)\neg(z \in X^1_{\mathbf{D}}),$$
- $\psi_2 =_{\text{Def}} ((\forall z)(z \in X^1_{\mathbf{I}} \leftrightarrow z \in X^1_{\top}) \wedge (\forall z)\neg(z \in X_{\perp}),$
- $\psi_3 =_{\text{Def}} \bigwedge_{A \in \mathsf{cpt}_{\mathcal{K}}} (\forall z)(z \in X^1_A \rightarrow z \in X^1_{\mathbf{I}}),$

- $\psi_4 =_{\text{Def}} (\bigwedge_{d \in N_D^{\mathcal{K}}} ((\forall z)(z \in X_d^1 \to z \in X_{\mathbf{D}}^1) \wedge \neg(\forall z)\neg(z \in X_d^1))$

$$\wedge (\forall z)(\bigwedge_{(d_i, d_j \in N_D^{\mathcal{K}}, i < j)} (z \in X_{d_i}^1 \leftrightarrow \neg(z \in X_{d_j}^1)))),$$

- $\psi_5 =_{\text{Def}} \bigwedge_{d \in N_D^{\mathcal{K}}} ((\forall z)(z \in X_d^1 \leftrightarrow z \in X_{\top_d}^1) \wedge (\forall z)\neg(z \in X_{\perp_d}^1)),$

- $\psi_6 =_{\text{Def}} \bigwedge_{d \in N_D^{\mathcal{K}}} \bigwedge_{f_d \in N_F^{\mathcal{K}}(d)} (\forall z)(z \in X_{f_d}^1 \to z \in X_d^1),$

- $\psi_7 =_{\text{Def}} (\forall z_1)(\forall z_2)((z_1 \in X_{\mathbf{I}}^1 \wedge z_2 \in X_{\mathbf{I}}^1) \leftrightarrow \langle z_1, z_2 \rangle \in X_U^3),$

- $\psi_8 =_{\text{Def}} \bigwedge_{R \in \text{arl}_{\mathcal{K}}} ((\forall Z^2)(Z^2 \in X_R^3 \to \neg(\forall z_1)(\forall z_2)\neg(\langle z_1, z_2 \rangle = Z^2))$

$$\wedge (\forall z_1)(\forall z_2)(\langle z_1, z_2 \rangle \in X_R^3 \to (z_1 \in X_{\mathbf{I}}^1 \wedge z_2 \in X_{\mathbf{I}}^1))),$$

- $\psi_9 =_{\text{Def}} \bigwedge_{T \in \text{crl}_{\mathcal{K}}} ((\forall Z^2)(Z^2 \in X_T^3 \to \neg(\forall z_1)(\forall z_2)\neg(\langle z_1, z_2 \rangle = Z^2))$

$$\wedge (\forall z_1)(\forall z_2)(\langle z_1, z_2 \rangle \in X_T^3 \to (z_1 \in X_{\mathbf{I}}^1 \wedge z_2 \in X_{\mathbf{D}}^1))),$$

- $\psi_{10} =_{\text{Def}} \bigwedge_{a \in \text{ind}_{\mathcal{K}}} (x_a \in X_{\mathbf{I}}^1) \wedge \bigwedge_{d \in N_D^{\mathcal{K}}} \bigwedge_{e_d \in N_C^{\mathcal{K}}(d)} x_{e_d} \in X_d^1,$

- $\psi_{11} =_{\text{Def}} \bigwedge_{e_{d_1}, \ldots, e_{d_n} \text{ in } \mathcal{K}} (\forall z)(z \in X_{\{e_{d_1}, \ldots, e_{d_n}\}}^1 \leftrightarrow \bigvee_{i=1}^n (z = x_{e_{d_i}}))$

$$\wedge \bigwedge_{a_1, \ldots, a_n \text{ in } \mathcal{K}} (\forall z)(z \in X_{\{a_1, \ldots, a_n\}}^1 \leftrightarrow \bigvee_{i=1}^n (z = x_{a_i})),$$

- $\psi_{12} =_{\text{Def}} \bigwedge_{d \in N_D^{\mathcal{K}}} \bigwedge_{\psi_d \in \text{bf}_{\mathcal{K}}^{\mathbf{D}}(d)} (\forall z)(z \in X_{\psi_d}^1 \leftrightarrow z \in \sigma(X_{\psi_d}^1)),$

with σ the transformation function from $4LQS^R$-variables of level 1 to $4LQS^R$-formulae recursively defined, for $d \in N_{\mathbf{D}}^{\mathcal{K}}$, by

$$\sigma(X_{\psi_d}^1) =_{\text{Def}} \begin{cases} X_{\psi_d}^1 & \text{if } \psi_d \in N_F^{\mathcal{K}}(d) \cup \{\top^d, \perp_d\} \\ \neg\sigma(X_{\chi_d}^1) & \text{if } \psi_d = \neg\chi_d \\ \sigma(X_{\chi_d}^1) \wedge \sigma(X_{\varphi_d}^1) & \text{if } \psi_d = \chi_d \wedge \varphi_d \\ \sigma(X_{\chi_d}^1) \vee \sigma(X_{\varphi_d}^1) & \text{if } \psi_d = \chi_d \vee \varphi_d. \end{cases}$$

In the above formulae, the variable $X_{\mathbf{I}}^1$ denotes the set of individuals $\mathbf{I}$, X_d^1 a datatype $d \in N_D^{\mathcal{K}}$, $X_{\mathbf{D}}^1$ a superset of the union of datatypes in $N_D^{\mathcal{K}}$, $X_{\top_d}^1$ and $X_{\perp_d}^1$ the constants $\top_d$ and $\perp_d$, and $X_{f_d}^1$, $X_{\psi_d}^1$ a facet f_d and a facet expression ψ_d, for $d \in N_D^{\mathcal{K}}$, respectively. In addition, X_A^1, X_R^3, X_T^3 denote a concept name A, an abstract role name R, and a concrete role name T occurring in $\mathcal{K}$, respectively. Finally, $X_{\{e_{d_1}, \ldots, e_{d_n}\}}^1$ denotes a data range $\{e_{d_1}, \ldots, e_{d_n}\}$ occurring in $\mathcal{K}$, and $X_{\{a_1, \ldots, a_n\}}^1$ a finite set $\{a_1, \ldots, a_n\}$ of nominals in $\mathcal{K}$.

Clearly, the constraints ψ_1-ψ_{12} have been introduced to guarantee that each model of $\varphi_{\mathcal{K}}$ can be easily transformed into a $\mathcal{DL}_{\mathbf{D}}^4$-interpretation.

Next we show that the consistency problem for $\mathcal{K}$ is equivalent to the satisfiability problem for $\varphi_{\mathcal{K}}$.

Let us first assume that $\varphi_{\mathcal{K}}$ is satisfiable. It is not hard to see that $\varphi_{\mathcal{K}}$ is satisfied by a $4LQS^R$-model of the form $\mathbf{M} = (D_1 \cup D_2, M)$, where:

- D_1 and D_2 are disjoint nonempty sets and $\bigcup_{d \in N_D^{\mathcal{K}}} d^{\mathbf{D}} \subseteq D_2$,

- $M X_{\mathbf{I}}^1 =_{\text{Def}} D_1$, $M X_{\mathbf{D}}^1 =_{\text{Def}} D_2$, $M X_d^1 =_{\text{Def}} d^{\mathbf{D}}$, for every $d \in N_D^{\mathcal{K}}$,

- $M X_{f_d}^1 =_{\text{Def}} f_d^{\mathbf{D}}$, for every $f_d \in N_F^{\mathcal{K}}(d)$, with $d \in N_D^{\mathcal{K}}$.

Exploiting the fact that $\mathcal{M}$ satisfies the constraints ψ_1-ψ_{12}, it is then possible to define a $\mathcal{DL}_\mathbf{D}^4$-interpretation $\mathbf{I}_\mathcal{M} = (\Delta^\mathbf{I}, \Delta_\mathbf{D}, \cdot^\mathbf{I})$, by putting $\Delta^\mathbf{I} =_{\mathrm{Def}} MX_\mathbf{I}^1$, $\Delta_\mathbf{D} =_{\mathrm{Def}} MX_\mathbf{D}^1$, $A^\mathbf{I} =_{\mathrm{Def}} MX_A^1$, for every concept name $A \in \mathsf{cpt}_\mathcal{K}$, $S^\mathbf{I} =_{\mathrm{Def}} MX_S^3$, for every abstract role name $S \in \mathsf{arl}_\mathcal{K}$, $T^\mathbf{I} =_{\mathrm{Def}} MX_T^3$, for every concrete role name $T \in \mathsf{crl}_\mathcal{K}$, and $a^\mathbf{I} =_{\mathrm{Def}} Mx_a$, for every individual $a \in \mathsf{ind}_\mathcal{K}$.

Since $\mathcal{M} \models \bigwedge_{H \in \mathcal{K}} \tau(H)$ and, as can be easily checked, $\mathbf{I}_\mathcal{M} \models_\mathbf{D} H$ if and only if $\mathcal{M} \models \tau(H)$, for every statement $H \in \mathcal{K}$, we plainly have $\mathbf{I}_\mathcal{M} \models_\mathbf{D} \mathcal{K}$, namely $\mathcal{K}$ is consistent, as we wished to prove.

Conversely, let $\mathcal{K}$ be a consistent $\mathcal{DL}_\mathbf{D}^4$-knowledge base. Then, there is a $\mathcal{DL}_\mathbf{D}^4$-interpretation $\mathbf{I} = (\Delta^\mathbf{I}, \Delta_\mathbf{D}, \cdot^\mathbf{I})$ such that $\mathbf{I} \models_\mathbf{D} \mathcal{K}$. We show how to construct, out of the datatype map $\mathbf{D}$ and the $\mathcal{DL}_\mathbf{D}^4$-interpretation $\mathbf{I}$, a $4LQS^R$-interpretation $\mathcal{M}_{\mathbf{I},\mathbf{D}} = (D_{\mathbf{I},\mathbf{D}}, M_{\mathbf{I},\mathbf{D}})$ which satisfies $\varphi_\mathcal{K}$. Let us put $D_{\mathbf{I},\mathbf{D}} =_{\mathrm{Def}} \Delta^\mathbf{I} \cup \Delta_\mathbf{D}$ and define $M_{\mathbf{I},\mathbf{D}}$ by putting $M_{\mathbf{I},\mathbf{D}} X_\mathbf{I}^1 =_{\mathrm{Def}} \Delta^\mathbf{I}$, $M_{\mathbf{I},\mathbf{D}} X_\mathbf{D}^1 =_{\mathrm{Def}} \Delta_\mathbf{D}$, $M_{\mathbf{I},\mathbf{D}} X_U^3 =_{\mathrm{Def}} U^\mathbf{I}$, $M_{\mathbf{I},\mathbf{D}} X_{dr}^1 =_{\mathrm{Def}} dr^\mathbf{D}$, for every variable X_{dr}^1 in φ denoting a data range dr occurring in $\mathcal{K}$, $M_{\mathbf{I},\mathbf{D}} X_A^1 =_{\mathrm{Def}} A^\mathbf{I}$, for every X_A^1 in φ denoting a concept name in $\mathcal{K}$, and $M_{\mathbf{I},\mathbf{D}} X_S^3 =_{\mathrm{Def}} S^\mathbf{I}$, for every X_S^3 in φ denoting an abstract role name in $\mathcal{K}$. Variable X_T^3, denoting concrete role names, and variables x_a, x_{e_d}, denoting individuals and datatype constants, respectively, are interpreted in a similar way. From the definitions of $\mathbf{D}$ and $\mathbf{I}$, it follows easily that $\mathcal{M}_{\mathbf{I},\mathbf{D}}$ satisfies the formulae ψ_1-ψ_{12} and $\tau(H)$, for every statement $H \in \mathcal{K}$, and, therefore, that $\mathcal{M}_{\mathbf{I},\mathbf{D}}$ is a model for $\varphi_\mathcal{K}$. $\qquad\square$

Some considerations on the expressive power of the logic $\mathcal{DL}_\mathbf{D}^4$ are in order. Despite $\mathcal{DL}_\mathbf{D}^4$ allows one to express existential quantification and at-least number restriction (resp., universal quantification and at-most number restriction) only on the left- (resp., right-) hand side of inclusion axioms, it is more liberal than $\mathcal{SROIQ}(\mathbf{D})$ in the construction of role inclusion axioms since the roles involved are not required to be subject to any ordering relationship. For example, the role hierarchy $\{RS \sqsubseteq S, RT \sqsubseteq R, VT \sqsubseteq T, VS \sqsubseteq V\}$ presented in [13] and not expressible in $\mathcal{SROIQ}(\mathbf{D})$ is admitted in the language of $\mathcal{DL}_\mathbf{D}^4$. Moreover, the notion of simple role is not needed in the definition of role inclusion axioms and of axioms involving number restrictions. Also, Boolean operators on roles are admitted and can be introduced in inclusion axioms such as, for instance, $R_1 \sqsubseteq R_2 \sqcap R_3$ and $R_1 \sqsubseteq \neg R_2 \sqcup R_3$. Finally, $\mathcal{DL}_\mathbf{D}^4$ treats derived datatypes by admitting datatype terms constructed from data ranges by means of a finite number of applications of the Boolean operators. Basic and derived datatypes can be used inside inclusion axioms involving concrete roles.

Remark 1. For a fixed positive integer h, a $\mathcal{DL}_\mathbf{D}^4$-knowledge base $\mathcal{K}$ is said to be *h-restricted* if an atom of any of the forms $R_1 \ldots R_{n_1} \sqsubseteq R$, $\geq_{n_2} R.C_1 \sqsubseteq C_2$, $\geq_{n_3} P.t_1 \sqsubseteq t_2$, $C_1 \sqsubseteq \leq_{n_4} R.C_2$, $t_1 \sqsubseteq \leq_{n_5} P.t_2$ occurs in $\mathcal{K}$ only if $n_1, n_2, n_3, n_4, n_5 \leq h$.

It turns out that by using the same function τ introduced in the proof of Theorem 1 and some additional constraints, the consistency problem for a h-restricted $\mathcal{DL}_\mathbf{D}^4$-knowledge base $\mathcal{K}$ can be expressed by a formula $\varphi'_\mathcal{K}$ such that

(i) $\varphi'_{\mathcal{K}}$ belongs to the sublanguage $(4LQS^R)^h$ of $4LQS^R$, whose satisfiability problem is **NP**-complete (see [8] for details), and

(ii) the size of $\varphi'_{\mathcal{K}}$ is polynomially related to that of $\mathcal{K}$.

From (i) and (ii) above, and from the **NP**-completeness of the satisfiability problem for propositional logic, it follows immediately that the consistency problem for h-restricted $\mathcal{DL}_{\mathbf{D}}^4$-knowledge bases is **NP**-complete.

Notice also that h-restricted $\mathcal{DL}_{\mathbf{D}}^4$-knowledge bases are quite expressive: for instance, in [18] we have shown that the ontology Ontoceramic, for ceramics classification, is representable in $(4LQS^R)^3$ and, much in the same way, it can be shown that it is representable as a 3-restricted $\mathcal{DL}_{\mathbf{D}}^4$-knowledge base.

3.1 Translating SWRL-rules into $4LQS^R$-formulae

The possibility of extending ontologies with rules has become a fundamental requirement to increase the expressiveness and the reasoning power of OWL knowledge bases. In a general sense, a rule is any sentence stating that if a set of premises is satisfied in a given model, then a certain conclusion must be satisfied in the same model. Although OWL is provided with several sorts of conditionals, these are, however, very constrained. Moreover, it is not possible to mix directly classes (concepts) and properties (roles) and include non-monotonic reasoning such as negation as failure.[3] Such considerations led to the definition of SWRL [20], a rule language combining OWL with the Unary/Binary Datalog fragment of the Rule Markup Language. SWRL allows users to write rules containing OWL constructs providing more reasoning capabilities than OWL alone.

An SWRL-rule r has the form $(\forall x_1, \ldots, x_n)(\mathsf{B} \implies \mathsf{H})$, where:

– B (the body of r) and H (the head of r) are conjunctions of atoms of the following types: $x \in C$, $y \in t$, $\langle x, y \rangle \in R$, $\langle x, y \rangle \in T$, $x = y$, $x \neq y$, with C a concept name, t a datatype, R an abstract role name, T a concrete role name, and x, y either individuals or variables (in the specific cases of atoms

Table 1. Examples of rule translation.

Type of Rule	Rule
SWRL-rule	$hasParent(X, Y), hasBrother(Y, Z) : -hasUncle(X, Z).$
$4LQS^R$-rule	$(\forall x)(\forall y)(\forall z)(\langle x, y \rangle \in X_{hasParent}^3 \wedge \langle y, z \rangle \in X_{hasBrother}^3 \rightarrow \langle x, z \rangle \in X_{hasUncle}^3)$
SWRL-rule	$Location(X), Trauma(Y), isLocationOf(X, Y), isPartOf(X, Z)$
	$:- isLocationOf(Z, Y)$
$4LQS^R$-rule	$(\forall x)(\forall y)(\forall z)(x \in X_{Location}^1 \wedge y \in X_{Trauma}^1 \wedge \langle x, z \rangle \in X_{isPartOf}^3 \rightarrow \langle z, y \rangle \in X_{isLocationOf}^3)$
SWRL-rule	$Person(X), hasAge(X, Y), (Y \geq 18) : -Adult(X)$
$4LQS^R$-rule	$(\forall x)(\forall y)(x \in X_{Person}^3 \wedge \langle x, y \rangle \in X_{hasAge}^3 \wedge y \in X_{\geq 18}^1 \rightarrow x \in X_{Adult}^1)$
SWRL-rule	$Region(Y), hasLocation(X, Y) : -hasRegion(X, Y)$
$4LQS^R$-rule	$(\forall x)(\forall y)(y \in X_{Region}^3 \wedge \langle x, y \rangle \in X_{hasLocation}^3 \rightarrow \langle x, y \rangle \in X_{hasRegion}^3)$

---

[3] We recall that a logic is non-monotonic if some conclusions can be invalidated when more knowledge is added.

of the forms $y \in t$ and $\langle x, y \rangle \in T$, y can be either a datatype constant or a variable), and

- $Var(\text{H}) \subseteq Var(\text{B}) = \{x_1, \ldots, x_n\}$, where $Var(\text{H})$ and $Var(\text{B})$ are the sets of variables occurring in H and in B, respectively.

In Table 1 we give some examples showing how SWRL-rules can be expressed by $4LQS^R$-formulae. For space reasons we do not provide here a formal translation function. However, it is not hard to see that it could be easily constructed by modifying the map τ introduced in the proof of Theorem 1.

4   Conclusions and Future Work

We have introduced the description logic $\mathcal{DL}_{\mathbf{D}}^4$ which admits, among other features, datatype reasoning, role chain axioms without regularity conditions on roles, min (resp., max) cardinality construct on the left-hand (resp., right-hand) side of inclusion axioms extended to non-simple roles, constructs of full negation, union, and intersection for abstract roles. As discussed at the end of Sect. 3, the logic $\mathcal{DL}_{\mathbf{D}}^4$ turns out to be quite expressive, if compared with $\mathcal{SROIQ}(\mathbf{D})$, the logic underpinning the Web Ontology Language OWL. However, although $\mathcal{DL}_{\mathbf{D}}^4$ is endowed with features not supported by $\mathcal{SROIQ}(\mathbf{D})$, it is not a proper extension of it, as $\mathcal{DL}_{\mathbf{D}}^4$ admits existential (resp., universal) quantification only on the left-hand (resp., right-hand) side of inclusion axioms.

Through a suitable translation process, we have then shown that the consistency problem for $\mathcal{DL}_{\mathbf{D}}^4$-knowledge bases can be effectively reduced to the satisfiability problem for the decidable fragment of set theory $4LQS^R$. Moreover, in the restricted case in which a $\mathcal{DL}_{\mathbf{D}}^4$-knowledge base $\mathcal{K}$ can involve only role chain axioms $R_1 \ldots R_m \sqsubseteq R$ and inclusion axioms $\geq_n R.C_1 \sqsubseteq C_2$, $C_1 \sqsubseteq \leq_p R.C_2$ such that m, n, and p do not exceed a fixed constant (hence independent of the size of $\mathcal{K}$), we have shown that the consistency problem is **NP**-complete, as it can be polynomially reduced to the satisfiability problem for a subfragment of $4LQS^R$, whose decision problem is **NP**-complete. As pointed out in Remark 1, such a restriction has a minimal impact on expressivity. Finally, we have also translated SWRL-rules into the $4LQS^R$ language.

We plan to introduce the constructs of union and intersection of concrete roles and to extend our results to include also datatype groups (here we have considered only a simple form of datatypes) and to admit Boolean operators on concrete roles by defining a suitable strategy of datatype checking. Moreover, we intend to extend the fragment $4LQS^R$ with metamodelling capabilities [16], so as to make it possible to define concepts containing other concepts and roles (i.e., meta-concepts) and relationships between concepts or between roles (i.e., meta-roles). We also plan to extend our complexity results, currently regarding only certain sublanguages of $4LQS^R$, concentrating on the complexity analysis of reasoning problems related to the whole $4LQS^R$ fragment. Finally, we intend to implement efficient reasoners for suitable fragments of $4LQS^R$.

References

1. Baader, F., Horrocks, I., Sattler, U.: Description logics as ontology languages for the semantic web. In: Hutter, D., Stephan, W. (eds.) Mechanizing Mathematical Reasoning. LNCS (LNAI), vol. 2605, pp. 228–248. Springer, Heidelberg (2005)
2. Cantone, D., Ferro, A., Omodeo, E.G.: Computable set theory. Number 6 in International Series of Monographs on Computer Science. Oxford Science Publications, Clarendon Press, Oxford, UK (1989)
3. Cantone, D., Longo, C.: A decidable two-sorted quantified fragment of set theory with ordered pairs and some undecidable extensions. Theor. Comput. Sci. **560**, 307–325 (2014)
4. Cantone, D., Longo, C., Nicolosi Asmundo, M.: A decidable quantified fragment of set theory involving ordered pairs with applications to description logics. In: Proceedings Computer Science Logic, 20th Annual Conference of the EACSL, CSL 2011, pp. 129–143, Bergen, Norway, 12–15 September 2011
5. Cantone, D., Longo, C., Nicolosi Asmundo, M.: A decision procedure for a two-sorted extension of multi-level syllogistic with the Cartesian product and some map constructs. In: Faber, W., Leone, N. (eds.) Proceedings of the 25th Italian Conference on Computational Logic (CILC 2010), Rende, Italy, 7–9 July 2010, vol. 598, pp. 1–18 (paper 11). CEUR Workshop Proceedings, ISSN 1613–0073, June 2010
6. Cantone, D., Longo, C., Nicolosi Asmundo, M., Santamaria, D.F.: Web ontology representation and reasoning via fragments of set theory (2015) CoRR, abs/1505.02075 (extended version)
7. Cantone, D., Longo, C., Pisasale, A.: Comparing description logics with multi-level syllogistics: the description logic $\mathcal{DL}\langle \mathcal{MLSS}_{2,m}^{\times}\rangle$. In: Traverso, P. (ed.) 6th Workshop on Semantic Web Applications and Perspectives, pp. 1–13, Bressanone, Italy, 21–22 September 2010
8. Cantone, D., Nicolosi-Asmundo, M.: Fundam. Inf. On the satisfiability problem for a 4-level quantified syllogistic and some applications to modal logic, vol. 124(4), pp. 427–448 (2013)
9. Cantone, D., Nicolosi-Asmundo, M., Santamaria, D.F., Trapani, F.: An ontology for ceramics cataloguing. In: Computer Applications and Quantitative Methods in Archaeology (CAA) (2015)
10. Cantone, D., Omodeo, E., Policriti, A.: Set Theory for Computing: From Decision Procedures to Declarative Programming with Sets. Monographs in Computer Science. Springer, New York (2001)
11. Dershowitz, N., Jouannaud, J.-P.: Rewrite systems. In: van Leeuwen, J. (ed.) Handbook of Theoretical Computer Science (vol. B), pp. 243–320. MIT Press, Cambridge, MA, USA (1990)
12. Formisano, A., Omodeo, E.G., Policriti, A.: Three-variable statements of set-pairing. Theor. Comput. Sci. **322**(1), 147–173 (2004)
13. Horrocks, I., Kutz, O., Sattler, U.: The even more irresistible SROIQ. In: Doherty, P., Mylopoulos, J., Welty, C.A. (eds.) Proceedings 10th International Conference on Principles of Knowledge Representation and Reasoning, pp. 57–67. AAAI Press (2006)
14. Horrocks, I., Sattler, U.: Ontology reasoning in the SHOQ(D) description logic. In: Proceeding of IJCAI 2001, pp. 199–204 (2001)
15. Kazakov, Y.: RIQ and SROIQ are harder than SHOIQ. In: Brewka, G., Lang, J. (eds.) Proceedings of the 11th International Conference, KR 2008, pp. 274–284, Sydney, Australia, 16–19 September 2008

16. Motik, B.: On the properties of metamodeling in OWL. In: Gil, Y., Motta, E., Benjamins, V.R., Musen, M.A. (eds.) ISWC 2005. LNCS, vol. 3729, pp. 548–562. Springer, Heidelberg (2005)

17. Motik, B., Horrocks, I.: OWL datatypes: design and implementation. In: Sheth, A.P., Staab, S., Dean, M., Paolucci, M., Maynard, D., Finin, T., Thirunarayan, K. (eds.) ISWC 2008. LNCS, vol. 5318, pp. 307–322. Springer, Heidelberg (2008)

18. Santamaria, D.F.: A Set-Theoretical Representation for OWL 2 Profiles. LAP Lambert Academic Publishing (2015). ISBN 978-3-659-68797-6

19. Schwartz, J.T., Cantone, D., Omodeo, E.G.: Computational Logic and Set Theory: Applying Formalized Logic to Analysis. Texts in Computer Science. Springer-Verlag New York Inc., New York (2011)

20. SWRL: http://www.w3.org/Submission/SWRL/

Allotment Problem in Travel Industry: A Solution Based on ASP

Carmine Dodaro[✉], Nicola Leone, Barbara Nardi, and Francesco Ricca

Department of Mathematics and Computer Science,
University of Calabria, Cosenza, Italy
{dodaro,leone,b.nardi,ricca}@mat.unical.it

Abstract. In the travel industry it is common for tour operators to
pre-book from service suppliers blocks of package tours, which are called
allotments in jargon. The selection of package tours is done according
to several preference criteria aimed at maximizing the expected earnings
given a budget. In this paper we formalize an allotment problem that
abstracts the requirements of a real travel agent, and we solve it using
Answer Set Programming. The obtained specification is executable, and
it implements an advanced feature of the iTravel+ system.

1 Introduction

In the travel industry it is common for tour operators to pre-book for the next
season blocks of package tours, which are called allotments in jargon [11,12]. This
practice is of help for both tour operators and service suppliers. Indeed, the first
have to handle possible market demand changes, whereas the seconds are subject
to the possibility that some package tours remain unsold, e.g., the rooms of a
hotel can remain empty in a given season. Therefore, service suppliers and tour
operators agree on sharing the economic risk of a potential low market demand
by signing allotment contracts [12]. The effectiveness of this form of supplying has
been studied in the economics literature under a number of assumptions on the
behavior of the contractors [11,21]. These studies, however, do not approach the
problem of providing a tool that helps travel agents in the act of selecting package
tours to be traded with service suppliers in the future market. Basically, given a
set of requirements on the properties of packages to be brought, budget limits,
and an offer of packages from several suppliers, the problem from the perspective
of the travel agent is to select a set of offers to be brought (or pre-booked) for the
next season so that the expected earnings are maximized [11]. Despite allotment
is –de facto– one of the most commonly-used supplying practices in the tourism
industry, the final selection of packages offered by travel suppliers is often done
in travel agencies more or less manually.

In this paper we approach the problem of automatic allotment of package
tours, and we formalize and solve it by using Answer Set Programming (ASP)

This work has been partially supported by the Calabrian Region under PIA project
iTravelPlus POR FESR Calabria 2007–2013.

© Springer International Publishing Switzerland 2015
B. ten Cate and A. Mileo (Eds.): RR 2015, LNCS 9209, pp. 77–92, 2015.
DOI: 10.1007/978-3-319-22002-4_7

[4,15,19]. ASP is a declarative rule-based programming paradigm for knowledge representation and declarative problem-solving. The idea of ASP is to represent a given computational problem by using a logic program, i.e., a set of logic ruler, such that its answer sets correspond to solutions, and then, use an answer set solver to find such solutions. The high knowledge-modeling power [4,15], and the availability of efficient systems [1,2,18], make ASP suitable for implementing complex knowledge-based applications. Indeed, ASP has been applied in several fields ranging from Artificial Intelligence [3,5] to Knowledge Management [4], and Information Integration [24]. In particular ASP has already been exploited in the tourism domain for identification of package tours that best suit the customers of an e-tourism system [27]. In this paper we follow this trend and we describe a new application of ASP in the touristic domain. The contributions of this paper can be summarized as follows:

- We abstract the requirements of a real travel agent that needs to solve an allotment problem, and we solve it by using an ASP program.
- We model in ASP a number of additional preference criteria on packages to be selected that, according to a travel agent advise, allow one to further optimize the selection process by taking into account additional knowledge of the domain.
- We report on the results of a preliminary experimental analysis using real-word data that validates our approach.

A tool based on the ASP encoding described in this paper will be included as an advanced service of the e-tourism platform developed under the iTravelPlus project by the Tour Operator Top Class s.r.l. and the University of Calabria.

The paper is structured as follows: Sect. 2 overviews ASP syntax and semantics; Sect. 3 exemplifies the requirements of an automatic allotment tool; Sect. 4 presents a formalization of the allotment problem in ASP; Sect. 5 presents the results of an empirical evaluation of our approach; Sect. 6 discusses related work, and Sect. 7 concludes the paper summarizing the obtained results.

2   Answer Set Programming

Answer Set Programming (ASP) [4,15,19] is a programming paradigm developed in the field of nonmonotonic reasoning and logic programming. In this section we overview the language of ASP, and we recall a methodology for solving complex problems with ASP. More detailed descriptions and a more formal account of ASP, including the features of the language employed in this paper, can be found in [8,16,19], whereas a nice introduction to ASP can be found in [4]. Hereafter, we assume the reader is familiar with logic programming conventions.

Syntax. The syntax of ASP is similar to the one of Prolog. Variables are strings starting with uppercase letter and constants are non-negative integers or strings starting with lowercase letters. A *term* is either a variable or a constant. A *standard atom* is an expression $p(t_1, \ldots, t_n)$, where p is a *predicate* of arity

n and $t_1, \ldots, t_n$ are terms. An atom $p(t_1, \ldots, t_n)$ is ground if $t_1, \ldots, t_n$ are constants. A *ground set* is a set of pairs of the form $\langle consts : conj \rangle$, where $consts$ is a list of constants and $conj$ is a conjunction of ground standard atoms. A *symbolic set* is a set specified syntactically as $\{Terms_1 : Conj_1; \cdots ; Terms_t : Conj_t\}$, where $t > 0$, and for all $i \in [1, t]$, each $Terms_i$ is a list of terms such that $|Terms_i| = k > 0$, and each $Conj_i$ is a conjunction of standard atoms. A *set term* is either a symbolic set or a ground set. Intuitively, a set term $\{X : a(X, c), p(X); Y : b(Y, m)\}$ stands for the union of two sets: The first one contains the X-values making the conjunction $a(X, c), p(X)$ true, and the second one contains the Y-values making the conjunction $b(Y, m)$ true. An *aggregate function* is of the form $f(S)$, where S is a set term, and f is an *aggregate function symbol*. Basically, aggregate functions map multisets of constants to a constant. The most common functions implemented in ASP systems are the following:

- `#min`, minimal term, undefined for the empty set;
- `#max`, maximal term, undefined for the empty set;
- `#count`, number of terms;
- `#sum`, sum of integers.

An *aggregate atom* is of the form $f(S) \prec T$, where $f(S)$ is an aggregate function, $\prec \in \{<, \leq, >, \geq\}$ is a comparison operator, and T is a term called guard. An aggregate atom $f(S) \prec T$ is ground if T is a constant and S is a ground set. An *atom* is either a standard atom or an aggregate atom. A *rule* r has the following form:

$$a_1 \mid \ldots \mid a_n \; :- \; b_1, \ldots, b_k, \text{not } b_{k+1}, \ldots, \text{not } b_m.$$

where $a_1, \ldots, a_n$ are standard atoms, $b_1, \ldots, b_k$ are atoms, $b_{k+1}, \ldots, b_m$ are standard atoms, and $n, k, m \geq 0$. A literal is either a standard atom a or its negation $\text{not } a$. The disjunction $a_1 \mid \ldots \mid a_n$ is the *head* of r, while the conjunction $b_1, \ldots, b_k, \text{not } b_{k+1}, \ldots, \text{not } b_m$ is its *body*. Rules with empty body are called *facts*. Rules with empty head are called *constraints*. A variable that appears uniquely in set terms of a rule r is said to be *local* in r, otherwise it is a *global* variable of r. An ASP program is a set of *safe* rules. A rule r is *safe* if both the following conditions hold: *(i)* for each global variable X of r there is a positive standard atom ℓ in the body of r such that X appears in ℓ; *(ii)* each local variable of r appearing in a symbolic set $\{Terms : Conj\}$ also appears in $Conj$.

A *weak constraint* [8] ω is of the form:

$$:\sim b_1, \ldots, b_k, \text{not } b_{k+1}, \ldots, \text{not } b_m. [w@l]$$

where w and l are the weight and level of ω. (Intuitively, $[w@l]$ is read "as weight w at level l", where weight is the "cost" of violating the condition in the body of ω, whereas levels can be specified for defining a priority among preference criteria). An ASP program with weak constraints is $\Pi = \langle P, W \rangle$, where P is a program and W is a set of weak constraints.

A standard atom, a literal, a rule, a program or a weak constraint is *ground* if no variables appear in it.

Semantics. Let P be an ASP program. The *Herbrand universe* U_P and the *Herbrand base* B_P of P are defined as usual (see e.g.,[4]). The ground instantiation G_P of P is the set of all the ground instances of rules of P that can be obtained by substituting variables with constants from U_P.

An *interpretation* I for P is a subset I of B_P. A ground literal ℓ (resp., not ℓ) is true w.r.t. I if $\ell \in I$ (resp., $\ell \notin I$), and false (resp., true) otherwise. An aggregate atom is true w.r.t. I if the evaluation of its aggregate function (i.e., the result of the application of f on the multiset S) with respect to I satisfies the guard; otherwise, it is false.

A ground rule r is *satisfied* by I if at least one atom in the head is true w.r.t. I whenever all conjuncts of the body of r are true w.r.t. I.

A model is an interpretation that satisfies all the rules of a program. Given a ground program G_P and an interpretation I, the *reduct* [16] of G_P w.r.t. I is the subset G_P^I of G_P obtained by deleting from G_P the rules in which a body literal is false w.r.t. I. An interpretation I for P is an *answer set* (or stable model [19]) for P if I is a minimal model (under subset inclusion) of G_P^I (i.e., I is a minimal model for G_P^I) [16].

Given a program with weak constraints $\Pi = \langle P, W \rangle$, the semantics of Π extends from the basic case defined above. Thus, let $G_\Pi = \langle G_P, G_W \rangle$ be the instantiation of Π; a constraint $\omega \in G_W$ is violated by an interpretation I if all the literals in ω are true w.r.t. I. An *optimum answer set* O for Π is an answer set of G_P that minimizes the sum of the weights of the violated weak constraints in G_W as a prioritized way.

Problem Solving in ASP. ASP can be used to encode problems in a declarative way. The disjunctive rules allow for expressing combinatorial problems which are in the second level of the polynomial hierarchy, and the (optional) separation of a fixed, non-ground program from an input database allows one to obtain uniform solutions over varying instances. More in detail, many problems of comparatively high computational complexity can be solved in a natural manner by following a *Guess&Check&Optimize* programming methodology [22]. This method requires that a database of facts is used to specify an instance of the problem; a set of (usually) disjunctive rules, called "guessing part", is used to define the search space; admissible solutions are then identified by other rules, called the "checking part", which impose some admissibility constraints; finally weak constraints are used to single out solutions that are optimal with respect to some criteria, the "optimize part". As an example, consider the well-known NP-hard problem called the Traveling Salesman Problem (TSP). Given a weighted graph $G = \langle V, E \rangle$, where V is the set of nodes and E is the set of edges with integer labels, the problem is to find a path of minimum length containing all the nodes of G. TSP can be encoded as follows:

$$vertex(v). \qquad \forall v \in V \tag{1}$$

$$edge(i, j, w). \quad \forall (i, j, w) \in E \tag{2}$$

$$inPath(X,Y) \mid outPath(X,Y) :- edge(X,Y,\_). \tag{3}$$

$$: - node(X), \#count\{I : inPath(I, X)\} \neq 1. \tag{4}$$
$$: - node(X), \#count\{O : inPath(X, O)\} \neq 1. \tag{5}$$
$$: - node(X), \text{not } reached(X). \tag{6}$$
$$reached(X) : - inPath(M, X), \#min\{N : node(N)\} = M. \tag{7}$$
$$reached(X) : - reached(Y), inPath(Y, X). \tag{8}$$
$$:\sim inPath(X, Y), edge(X, Y, W).[W@0] \tag{9}$$

The first two lines introduce suitable facts, representing the input graph G. Then, rule (3), which can be read as "each edge may or may not be part of the path", guesses a solution (a set of inPath atoms). Lines 4–6 select admissible paths. In particular, rule in line (4) (resp., (5)) is satisfied if exactly one arc in the solution enters in (resp., exits from) each node, indeed the body is satisfied by solutions not having a count of one. Moreover, line (6) ensures that the path traverses (say, reaches) all the nodes of G. Actually, this latter condition is obtained by checking that there exists a path reaching all the nodes of G and starting from the first node of V, say M. In particular, a node X is reached either if there is an arc connecting M to X (rule (7)), or if there is an arc connecting a reached node Y to X (rule (8)). The last line selects the solutions of minimal weight. Indeed the weak constraint in rule (9) is violated with cost W each time an arc of cost W is in the candidate solution.

3   Travel Agent Requirements

In this section we informally describe the requirements of a common problem in the tourism industry, i.e. the problem of booking in advance blocks of package tours for the next season. A travel agency usually selects a block of package tours from several travel suppliers, which may apply several discounts if predetermined amounts of their package tours are bought. In general, a critical requirement is that *the sum of prices of selected package tours must not exceed a limited budget.* This means that travel agencies are not allowed to buy all the package tours they wish. Thus, their goal is to *select package tours in order to maximize the expected earnings.* Moreover, depending on the specific needs, travel agencies might specify other preferences among the selected package tours. Those preferences are not general. On the contrary, two different travel agencies usually have different priorities among selected packages according to their experience and customer base. In the following we detail several preferences that travel agencies might specify according to their needs.

Preference of Suppliers According to Destination. Travel agencies might specify a preference of suppliers for package tours involving particular destinations. For instance, a supplier can be considered highly reliable for travels in Europe and unreliable for travels in other countries.

Preference of Suppliers According to the Type of Holidays. A supplier can be considered also preferable for particular types of holidays. For instance, some suppliers are specialized in holidays involving cruises, while others are specialized in holidays involving sports activities.

Preference on Package Tours with the Highest Rating. After a trip, travelers usually evaluate their holidays by assigning a numerical score. A package tour is evaluated by looking at the rating assigned by the travelers. Thus, travel agencies give priority to the package tours with the highest ratings.

Preferences on the Number of Package Tours to Buy. According to their typology of customers travel agencies also express a preference on the number of package tours to buy. In particular, in case travel agencies obtain the same expected earnings from two or more package tours then they can maximize or minimize the number of bought package tours according to their customer base. For instance, travel agencies working with wealthy customers may prefer to buy few package tours with highest earnings, while travel agencies working with many customers may prefer to maximize the number of package tours to buy.

Preference on the Amount of Money to Pay. Another important preference concerns the amount of money to pay. In particular, in case travel agencies obtain the same expected earnings from two or more package tours it is preferable to select package tours with the lowest prices.

4   Specification in ASP

This section illustrates the ASP program which solves the allotment problem specified in the previous section. First, the input data is described, then, the ASP rules solving the allotment problem are presented. Finally, preferences that can be specified by travel agencies are described.

4.1   Data Model

The input of the process is specified by means of the predicates described in this section. The predicates representing the facts of our encoding are the following:

- Instances of the predicate *availablePackage(pkId, supplier, destination, type, sellingPrice, purchasePrice, rating, availableQuantity)* represent stocks of available package tours in the market, where *pkId* is the identifier of the tour package, *supplier* is the identifier of the supplier selling the package tour, *destination* is the destination of the package tour, *type* is the type of holiday, *sellingPrice* is the price applied by the travel agency to their customers, *purchasePrice* is the price applied by the supplier to the travel agency, *rating* is a numerical score associated to the package tour representing the appreciation of customers for this package tour, and *availableQuantity* corresponds to the quantity of available package tours of this kind in the market.

- Instances of the predicate *requiredPackage(destination, type, minPrice, max-Price, requiredQuantity)* represent package tours required by a travel agency, where *destination* is destination of the package tour required, *type* is the type of holiday required, *minPrice* and *maxPrice* represents the range of prices the travel agency is willing to pay for a given destination and type of holiday, and *requiredQuantity* corresponds to the quantity of required package tours of this kind.
- Instances of the predicate *discount(supplier, quantity, percentageDiscount)* represent the discount applied by suppliers if a given amount of their package tours is bought, where *supplier* represents the identifier of the supplier, *quantity* is the minimum quantity of bought package tours for applying the discount, and *percentageDiscount* is the percentage discount applied.
- The only instance of the predicate *budget(b)* represents the maximum amount of money the travel agency is willing to pay.
- Instances of the predicate *evalSupplierDestination(supplier, destination, score)* represent the evaluations of suppliers according to destination, where *supplier* is the identifier of the supplier, *destination* is the destination of the package tour, and *score* is a numerical score representing the reliability of the supplier for package tours involving the destination.
- Instances of the predicate *evalSupplierType(supplier, type, score)* represent the evaluations of suppliers according to type of holiday, where *supplier* is the identifier of the supplier, *type* is the type of holiday, and *score* is a numerical score representing the reliability of the supplier for package tours concerning the type of holiday.

4.2   Allotment Problem

In this section we describe the ASP rules used for solving the allotment problem. We follow the *Guess&Check&Optimize* programming methodology [22]. In particular, the following disjunctive rule guesses a quantity to buy for each required package:

$$buy(P, Q) \mid nBuy(P, Q) : - availablePackages(P, \_, D, T, SP, PP, \_, AvQ),$$
$$requiredPackages(D, T, MinP, MaxP, ReqQ),$$
$$0 \le Q \le ReqQ,$$
$$Q \le AvQ,$$
$$MinP \le SP \le MaxP.$$

$$(1)$$

The guess of the quantity is limited to available package tours which are requested and their selling price is in the requested range. Then, assignments buying different quantities of the same package tour are filtered out by the following constraint:

$$: - \#count\{Q, P : buy(P, Q)\} > 1, availablePackages(P, \_, \_, \_, \_, \_, \_, \_). \quad (2)$$

Suppliers may apply one or more discounts if predetermined amounts of their package tours are bought. In general several discounts are offered depending on the volume of booked packages. In this case the maximum applicable discount among them must be applied. All applicable discounts and the maximum discount among them are computed by the following rules:

$$allDiscounts(S, D) : -discount(S, Q1, D),$$
$$\#sum\{Q, P : buy(P, Q)\} \geq Q1. \tag{3}$$

$$maxDiscount(S, Disc) : -discount(S, \_, \_),$$
$$\#max\{D : allDiscounts(S, D)\} = Disc.$$

The predicate *allDiscounts(supplier, discount)* stores the association between the supplier and all the applicable discounts, while *maxDiscount(supplier, discount)* stores the association between the supplier and the corresponding maximum applicable discount. Then, the prices of the package tours are updated according to the above-computed discounts. This behavior is achieved by employing the following rule:

$$discountPrices(P, SP, PPD) : -availablePackages(P, S, \_, \_, SP, PP, \_, \_),$$
$$maxDiscount(S, MD),$$
$$PPD = PP - (PP * MD)/100. \tag{4}$$

The predicate *discountPrices* stores the original selling price and the purchase price after the application of the discount for each package tour. This predicate is then used to handle a critical requirement on the budget, i.e. the sum of prices of selected package tours must not exceed a limited budget. This is expressed in ASP by the following rule:

$$: -\#sum\{PP * Q, P : buy(P, Q), discountPrices(P, \_, PP)\} > B,$$
$$budget(B). \tag{5}$$

Finally, the last requirement is to maximize the earnings. This is obtained in our encoding by means of the following weak constraint:

$$:\sim discountPrices(P, SP, PP), buy(P, Q),$$
$$E = (SP - PP) * Q.[-E@\ell] \tag{6}$$

Intuitively, when a stock of package tours is bought the solution is associated with a cost depending on the earnings obtained by buying those packages. The weight of weak constraint is negative since weak constraints expresses the minimization of the cost associated to a solution.[1] The choice of the level ℓ is explained in the following section.

---

[1] ASP solvers may have undefined behaviors in presence of negative weights. A workaround is to augment the weight of the weak constraint by the maximum possible earnings.

4.3 Preferences (optional)

In this section, we describe preferences travel agencies might specify among the selected package tours depending on their specific needs. Different travel agencies usually have different priorities among selected package tours which are expressed in our framework by means of weak constraints. In the weak constraints we use numerical values $\ell_1, \ldots, \ell_5$ representing the levels of weak constraints. Then, an order on the preferences can be specified by properly assigning a value to those levels. The only requirement is that the level ℓ of the constraint that maximizes earnings (6) is greater than all the other weak constraints that are specified in the following.

Preference of Suppliers According to Destination. A travel agency might specify a preference of suppliers according to the destination of a travel. The following weak constraint expresses this preference:

$$:\sim evalSupplierDestination(S, D, SC),$$
$$availablePackages(P, S, D, \_, \_, \_, \_, \_), \tag{7}$$
$$nBuy(P, Q).[SC * Q@\ell_1]$$

Intuitively, when a stock of package tours is not bought a numerical penalty is associated to the solution. For each package tour which is not selected the cost of the solution is increased by the score associated to the corresponding supplier for the destination.

Preference of Suppliers According to the Type of Holidays. Similarly, the following weak constraint expresses a preference among suppliers according to the type of holidays:

$$:\sim evalSupplierType(S, T, SC),$$
$$availablePackages(P, S, \_, T, \_, \_, \_, \_), \tag{8}$$
$$nBuy(P, Q).[SC * Q@\ell_2]$$

For each package that is not selected the solution cost is increased by the score associated to the corresponding supplier for the type of holiday. The effect is to maximize the number of package tours in the solution that are provided by preferred suppliers.

Preference on Package Tours with the Highest Rating. Travel agencies give priority to the package tours with the highest ratings. This preference is expressed by the following weak constraint:

$$:\sim availablePackages(P, \_, \_, \_, \_, \_, R, \_), nBuy(P, Q).[R * Q@\ell_3] \tag{9}$$

Here, the cost of the solution is given by the sum of ratings of package tours which are not bought. Thus we maximize the ratings of selected package tours.

Preferences on the Number of Package Tours to Buy. In case a travel agency is willing to minimize the number of packages to buy we apply the following weak constraint:

$$:\sim buy(P, Q).[Q@\ell_4] \tag{10}$$

The cost of the solution is increased by the quantity of package tours which are not bought. Otherwise, if a travel agency is willing to maximize the number of packages to buy we apply the following weak constraint:

$$:\sim nBuy(P, Q).[Q@\ell_4] \tag{11}$$

Here, the cost of the solution is increased by the quantity of package tours which are bought. Note that weak constraints (10) and (11) are never applied together since travel agencies either maximize or minimize the number of package tours to buy.

Preference on the Amount of Money to Pay. Finally, travel agencies may also want to minimize the amount of money to pay. Note that this is different from the earnings, since in this case travel agency minimizes the purchase prices without considering their selling prices. This behavior is employed by the following weak constraint:

$$:\sim discountPrices(P, \_, PP), buy(P, Q).[PP * Q@\ell_5] \tag{12}$$

Intuitively, the cost of the solution depends on sum of prices of package tours which are bought. Hence, this has the effect to minimize the price of package tours in the solution.

Specification of Preferences. As stated in Sect. 3, the preferences depend on the specific needs of travel agencies, and can be applied selectively by simply adding or ignoring some of the weak constraints described in Sect. 4.3. Moreover, a travel agent must also specify a layering of preferences by properly assigning values to $\ell_1, \ldots, \ell_5$. As an example, consider a travel agent that wants to give highest priorities on package tours with the highest ratings; and then maximize the number of packages to buy. In the encoding those preferences are specified by considering weak constraints (9) and (11) and by assigning integer values to the levels such that $\ell_3 > \ell_5$, e.g., $\ell_3 = 2$, and $\ell_5 = 1$.

5   Empirical Validation

We validated our ASP-based solution running a preliminary experiment on real-world data provided by the partners of the iTravelPlus project. In particular, we obtained an instance of a database of package tours querying the database of the iTravel+ system and properly encoding it by means of ASP facts. Moreover, we generated a specification of the requested package tours by running a mining services of the same system that generates a prediction based on the package tours

sold in the past. Finally, we randomly generated a number of additional requirements to test the effects of the optional preferences of our solution. Concerning
the ASP solver we used WASP [2]. (For the sake of completeness, we report that
we also tried the ASP solver CLASP [18] obtaining similar performance.) The
system was run on a four core Intel Xeon CPU X3430 2.4 GHz, with 16 GB of
physical RAM, each execution was limited to 600 seconds.

Table 1. Performance of the system for different available package tours.

	Available Pkgs (Min-Max)	#inst	#solved	#optima	Time (no pref.)	Time (all pref.)
DBx0.5	216–291	30	30	30	0.6	0.8
DBx1	445–584	30	30	26	48.5	147.8
DBx2	963–1093	30	30	14	11.7	33.41

Table 2. Performance of the system for different periods (in months).

Period	Required Pkgs (Min-Max)	#inst	#solved	#optima	Search Space (avg)
2m	79–94	30	30	30	2^{27}
4m	174–182	30	30	24	2^{54}
6m	345–365	30	30	16	2^{110}

The performance of the system for different sizes of the available packages
is reported in Table 1. In particular, the first column reports the sizes of the
considered DBs. We considered the original database (labeled DBx1) and then we
consider two more settings containing the first half of the same database (labeled
DBx0.5), a generated instance (labeled DBx2) having twice of the facts from
DBx1 obtained adding more suppliers. The second column reports the minimum
and the maximum available package tours among the instances considered. The
number of considered instances is reported in the third column together with the
number of instances in which the system found a (sub-optimal) solution (fourth
column) and the number of instances in which the system found the optimum
solution (fifth column). The sixth column reports the sum of execution times
(in seconds) elapsed for finding the optimum solution of the allotment problem
without optional preferences. The seventh column reports the sum of time in
seconds of finding the optimum solution of the allotment problem where all
preferences are enabled. As first observation we note that the system provides a
solution for all the instances considered within 10 min. The provided solutions
are usually either close to the optimum ones or are optimal but the system is not
able to prove their optimality within the timeout. Moreover, when we consider
half size of the DB size the system finds always the optimum solution, while
for the original size of the DB this is the case in the 87 % of the instances,
which is a performance considered fairly acceptable by our project partners.
The performance is still good in the case we double the size of the original DB,

since the system is able to find the optimum for about half of the instances. In addition, we also observe that adding the preference does not reduce either the number of solved instances nor the number of instances in which the optimum solution is found. As expected, we observe a constant slow down in the solving time, which is approximately three times higher than the one measured with no preferences.

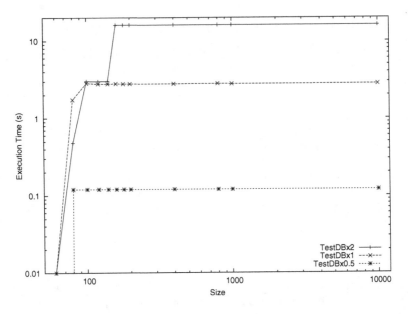

Fig. 1. Scalability w.r.t. the number of available package tours.

It is worth pointing out that the performance of the system does not heavily depend on the quantity of available packages. In fact, rule (1) in Sect. 4.2 filters out all the package tours which are not required. In order to confirm this observation, we increased the available quantities for each package tours in the database by several factors. The result is reported in Fig. 1, in which, for a particular size of the DB, a point (x, y) represents the solving time y where the availability of package tours is x percent of the original size. The graph shows that the execution times grow until the offered packages are 120 % more than in the original DB, and then performance has a constant trend that is not dependent on the quantity of available packages.

Table 2 reports on the performance of the system for different periods of request packages. In particular, the first column reports the considered period expressed in months. The second column reports the minimum and the maximum required package tours among the instances considered. The number of considered instances is reported in the third column together with the number of instances in which the system found a solution (fourth column) and the number of instances in which the system found the optimum solution (fifth column).

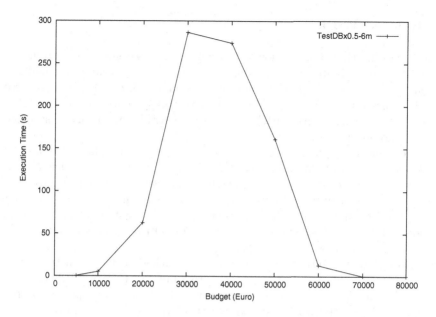

Fig. 2. Performance of the system depending on the budget.

The last column reports the average search space for the considered instances. Also in this case, the system provides a solution for all the instances considered. Moreover, when we consider a period of 2 months the system also finds always the optimum solution, while if we considered a period of 4 months this is the case in the 80 % of the instances. It is worth pointing out that, travel agencies usually book package tours for one season, thus they consider a period of at most 3–4 months. Nonetheless, the performance is still good in the case we consider a period of 6 months, since the system is able to find the optimum for about half of the instances.

Finally, another observation concerns the budget allowed by the travel agency, since the hardness of the instances depends on this parameter. In fact, is it easy to see that instances with very low (resp. high) budgets w.r.t. to the one needed to fulfill the request are likely easy, since they correspond to over-constrained (resp. under-constrained) problems where the solution is to buy no package tours (resp. to buy all required package tours). Thus, we also analyzed the behavior of the system in case we consider different budgets. The result is reported for DBx0.5 and a request of 6 months in Fig. 2, in which a point (x, y) represents the solving time y if the budget is limited to x euro. The trend of the system confirms our expectations, since the instance is trivially solved when the budget is enough either to buy nothing or to buy everything. The maximum hardness is reached when the allotted budget can cover about 40 % of the request in our experiment, a setting that in real-world instances is not that common, since the budget is usually enough to cover most of the requests.

6 Related Work

In the literature there are solution to many e-tourism systems challenges including: package tours search and assemblage, automatic holiday advisors, modeling of general purpose ontologies of the touristic domain [6,9,10,14,23,25,26], etc. These studies do not focus –to the best of our knowledge– on helping travel agents in the act of selecting package tours to be traded with service suppliers in the future market.

Concerning the applications of ASP, we mention that it has been used to develop several industrial applications [20,28] and, in particular, it has already been exploited in an e-tourism system [27]. Nonetheless, the problem considered in [27] was to identify the package tours that best suit the needs of a customer of an e-tourism platform; thus, [27] approaches a different problem of the one considered here. This paper presents the first attempt to exploit ASP for assisting tour operators in the allotment of packages.

It is important pointing out that other approaches, such as Constraint Programming (CP) [29] and Mixed-Integer Programming (MIP) [13], could also be good candidates for solving the problem considered in this paper. Despite it would be interesting to investigate whether other solving technologies such as CP or MIP are also successfully applicable, the goal of this research was to demonstrate that ASP can be used for solving the allotment problem in practice, and our ASP-based solution satisfied the stakeholder partners of the iTravelPlus project. Comparison with other approaches can be an interesting research goal to be developed in a separate work.

For the sake of completeness, we also mention a different way of dealing with the problem of allotment [30]. This approach aims at acquiring directly and on-demand from hotel management services the information about the hotel rooms and facilities that suit the request of a tour operator, so to avoid the allotment problem using agent technologies [30]. This is clearly a radically different approach from ours that aims at optimizing the pre-booking of allotments for an entire period of time.

7 Conclusion

In this paper we described an application of Answer Set Programming to the problem of allotment in travel industry. We have formalized an allotment problem that abstracts the requirements of a real travel agent, by means of an ASP programs. Since ASP programs are executable specifications we also obtained a prototypical implementation of a tool for supporting a travel agent in selecting the packages to be traded for next season. We experimented with our implementation on instances of the problem made of real-word data provided by the travel agency Top Class s.r.l. The preliminary results that we obtained are promising in terms of performance. Our ASP program will be included as an advanced reasoning service of the e-tourism platform developed under the iTravelPlus project by the Tour Operator Top Class s.r.l. and the University of Calabria.

As far as future work is concerned, we plan to study the computational properties of the allotment problem, and to extend our formulation by modeling additional preference criteria on the solutions. In this respect we will investigate the adoption of more general frameworks for expressing preferences [7,17]. The automatic allotment tool will be integrated in the system developed under the iTravelPlus project that aims at developing several services for tour operators.

Acknowledgments. We are grateful to DLV SYSTEM s.r.l. for its support in the development of the system and to Denise Angilica, Gianluigi Greco, and Gianni Laboccetta for fruitful discussions on the specification of the problem.

References

1. Alviano, M., Calimeri, F., Charwat, G., Dao-Tran, M., Dodaro, C., Ianni, G., Krennwallner, T., Kronegger, M., Oetsch, J., Pfandler, A., Pührer, J., Redl, C., Ricca, F., Schneider, P., Schwengerer, M., Spendier, L.K., Wallner, J.P., Xiao, G.: The fourth answer set programming competition: preliminary report. In: Cabalar, P., Son, T.C. (eds.) LPNMR 2013. LNCS, vol. 8148, pp. 42–53. Springer, Heidelberg (2013)
2. Alviano, M., Dodaro, C., Faber, W., Leone, N., Ricca, F.: WASP: a native asp solver based on constraint learning. In: Cabalar, P., Son, T.C. (eds.) LPNMR 2013. LNCS, vol. 8148, pp. 54–66. Springer, Heidelberg (2013)
3. Balduccini, M., Gelfond, M., Watson, R., Nogueira, M.: The USA-Advisor: a case study in answer set planning. In: Eiter, T., Faber, W., Truszczyński, M. (eds.) LPNMR 2001. LNCS (LNAI), vol. 2173, p. 439. Springer, Heidelberg (2001)
4. Baral, C.: Knowledge Representation, Reasoning and Declarative Problem Solving. Cambridge University Press, Cambridge (2003)
5. Baral, C., Gelfond, M.: Reasoning agents in dynamic domains. In: Minker, J. (ed.) Logic-Based Artificial Intelligence, pp. 257–279. Kluwer Academic Publishers, Dordrecht (2000)
6. Barta, R., Feilmayr, C., Pröll, B., Grün, C., Werthner, H.: Covering the semantic space of tourism: an approach based on modularized ontologies. In: CIAO, pp. 1–8. ACM (2009)
7. Brewka, G., Delgrande, J.P., Romero, J., Schaub, T.: asprin: Customizing answer set preferences without a headache. In: Bonet, B., Koenig, S. (eds.) AAAI. pp. 1467–1474. AAAI Press (2015)
8. Buccafurri, F., Leone, N., Rullo, P.: Enhancing disjunctive datalog by constraints. IEEE Trans. Knowl. Data Eng. **12**(5), 845–860 (2000)
9. Cardoso, J.: Combining the semantic web with dynamic packaging systems. In: AIKED. pp. 133–138. World Scientific and Engineering Academy and Society (2006)
10. Cardoso, J.: Developing an owl ontology for e-tourism. In: Cardoso, J., Sheth, A.P. (eds.) Semantic Web Services, Processes and Applications, pp. 247–282. Springer, Heidelberg (2006)
11. Castellani, M., Mussoni, M.: An economic analysis of tourism contracts: allotment and free sale*. In: Matias, A., Nijkamp, P., Neto, P. (eds.) Advances in Modern Tourism Research, pp. 51–85. Springer, Heidelberg (2007)

12. Cooper, C., Fletcher, J., Fyall, A., Gilbert, D., Wanhill, S.: Tourism: Principles and Practice, 4 pap/pas edn. Financial Times Management, Harlow (2008)
13. Dakin, R.J.: A tree-search algorithm for mixed integer programming problems. Comput. J. **8**(3), 250–255 (1965)
14. Dogac, A., Kabak, Y., Laleci, G., Sinir, S., Yildiz, A., Kirbas, S., Gurcan, Y.: Semantically enriched web services for the travel industry. SIGMOD Rec. **33**(3), 21–27 (2004)
15. Eiter, T., Gottlob, G., Mannila, H.: Disjunctive datalog. ACM Trans. Database Syst. **22**(3), 364–418 (1997)
16. Faber, W., Leone, N., Pfeifer, G.: Semantics and complexity of recursive aggregates in answer set programming. Artif. Intell. **175**(1), 278–298 (2008)
17. Gebser, M., Kaminski, R., Kaufmann, B., Schaub, T.: Multi-criteria optimization in answer set programming. In: Gallagher, J.P., Gelfond, M. (eds.) Technical Communications of ICLP, LIPIcs, vol. 11, pp. 1–10. Schloss Dagstuhl - Leibniz-Zentrum fuer Informatik, Wadern (2011)
18. Gebser, M., Kaufmann, B., Schaub, T.: Conflict-driven answer set solving: from theory to practice. Artif. Intell. **187**, 52–89 (2012)
19. Gelfond, M., Lifschitz, V.: Classical negation in logic programs and disjunctive databases. New Gener. Comput. **9**, 365–385 (1991)
20. Grasso, G., Leone, N., Manna, M., Ricca, F.: Logic Programming, Knowledge Representation, and Nonmonotonic Reasoning: Essays in Honor of Michael Gelfond. LNAI, vol. 6565. Springer, Heidelberg (2011)
21. Gurcaylilar-Yenidogan, T., Yenidogan, A., Windspergerc, J.: Antecedents of contractual completeness: the case of tour operator-hotel allotment contracts. Procedia Soc. Beha. Sci. **24**, 1036–1048 (2011). International Strategic Management Conference
22. Leone, N., Pfeifer, G., Faber, W., Eiter, T., Gottlob, G., Perri, S., Scarcello, F.: The DLV system for knowledge representation and reasoning. ACM Trans. Comput. Log. **7**(3), 499–562 (2006)
23. Maedche, A., Staab, S.: Applying semantic web technologies for tourism information systems. In: Weber, K., Frew, A., Hitz, M. (eds.) International Conference for Information and Communication Technologies in Tourism. Springer, Heidelberg (2002)
24. Manna, M., Ricca, F., Terracina, G.: Consistent query answering via ASP from different perspectives: theory and practice. TPLP **13**(2), 252–277 (2013)
25. Martin, H., Katharina, S., Daniel, B.: Towards the semantic web in e-tourism: can annotation do the trick? In: European Conference on Information System (2006)
26. Prantner, K., Ding, Y., Luger, M., Yan, Z., Herzog, C.: Tourism ontology and semantic management system: State-of-the-arts analysis. In: International Conference WWW/Internet. IADIS (2007)
27. Ricca, F., Dimasi, A., Grasso, G., Ielpa, S.M., Iiritano, S., Manna, M., Leone, N.: A logic-based system for e-tourism. Fundam. Inform. **105**(1–2), 35–55 (2010)
28. Ricca, F., Grasso, G., Alviano, M., Manna, M., Lio, V., Iiritano, S., Leone, N.: Team-building with answer set programming in the gioia-tauro seaport. TPLP **12**(3), 361–381 (2012)
29. Rossi, F., Beek, P.v., Walsh, T.: Handbook of Constraint Programming (Foundations of Artificial Intelligence). Elsevier Science Inc., New York, NY, USA (2006)
30. Withalm, J., Karl, E., Fasching, M.: Agents solving strategic problems in tourism. In: Fesenmaier, D., Klein, S., Buhalis, D. (eds.) Information and Communication Technologies in Tourism 2000, pp. 275–282. Springer (2000)

A Rule-based Framework for Creating Instance Data from *OpenStreetMap*

Thomas Eiter[1], Jeff Z. Pan[3], Patrik Schneider[1,4](✉), Mantas Šimkus[1], and Guohui Xiao[2]

[1] Institute of Information Systems, Vienna University of Technology, Vienna, Austria
patrik@kr.tuwien.ac.at, simkus@dbai.tuwien.ac.at
[2] Faculty of Computer Science, Free University of Bozen-Bolzano, Bolzano, Italy
[3] University of Aberdeen, Aberdeen, UK
[4] Vienna University of Economics and Business, Vienna, Austria

Abstract. Reasoning engines for ontological and rule-based knowledge bases are becoming increasingly important in areas like the Semantic Web or information integration. It has been acknowledged however that judging the performance of such reasoners and their underlying algorithms is difficult due to the lack of publicly available datasets with large amounts of *(real-life) instance data*. In this paper we describe a framework and a toolbox for creating such datasets, which is based on extracting instances from the publicly available *OpenStreetMap* (*OSM*) geospatial database. To this end, we give a formalization of OSM and present a rule-based language to specify the rules to extract instance data from OSM data. The declarative nature of the approach in combination with external functions and parameters allows one to create several variants of the dataset via small modifications of the specification. We describe a highly flexible toolbox to extract instance data from a given OSM map and a given set of rules. We have employed our tools to create benchmarks that have already been fruitfully used in practice.

1 Introduction

Reasoning over ontological and rule-based knowledge bases (KBs) is receiving increasing attention. In particular *Description Logics* (DLs), which provide the logical foundations to OWL ontology languages, are a well-established family of decidable logics for knowledge representation and reasoning. They offer a range of expressivity well-aligned with computational complexity. Moreover, several systems have been developed in the last decade to reason over DL KBs, which usually consist of a *TBox* that describes the domain in terms of *concepts* and *roles*, and an *ABox* that stores information about known *instances* of concepts and their participation in roles.

Supported by the Vienna Science and Technology Fund (WWTF) project ICT12-15, by the Austrian Science Fund (FWF) project P25207, and by the EU project Optique FP7-318338.

B. ten Cate and A. Mileo (Eds.): RR 2015, LNCS 9209, pp. 93–104, 2015.
DOI: 10.1007/978-3-319-22002-4_8

Naturally, classical reasoning tasks like TBox satisfiability and subsumption under a TBox have received most attention and many reasoners have been devoted to them. A different category are reasoners for *ontology-based query answering* (*OQA*), which are designed to answer queries over DL KBs in the presence of large data instances (see e.g. Ontop [14], Pellet [19], and OWL-BGP [13]). TBoxes in this setting are usually expressed in low complexity DLs, and are relatively small in size compared to the instance data. These features make reasoners for OQA different from classical (TBox) reasoners. The DL community is aware that judging the performance of OQA reasoners and their underlying algorithms is difficult due to the lack of publicly available benchmarks consisting of large amounts of *real-life instance data*. In particular, the popular Lehigh University Benchmark (LUBM) [10] only allows one to generate random instance data, which provides only a limited insight into the performance of OQA systems.

In this paper, we consider publicly available geographic datasets as a source of test data for OQA systems and other types of reasoners. For the benchmark creation, we need a framework and a toolbox for extracting and enhancing instance data from *OpenStreetMap* (*OSM*) geospatial data.[1] The OSM project aims to collaboratively create an open map of the world. It has proven hugely successful and the map is constantly updated and extended. OSM data describes maps in terms of (possibly *tagged*) points, geometries, and more complex aggregate objects called *relations*. We believe the following features make OSM a good source to obtain instance data for reasoners: (a) Datasets of different sizes exist; e.g., OSM maps for all major cities, countries, and continents are directly available or can be easily generated. (b) Depending on the location (e.g., urban versus rural), the density, separation, and compactness of object location varies strongly.[2] (c) Spatial objects have an inherent structure of containment, bordering, and overlapping, which can be exploited to generate spatial relations (e.g., contains). (d) Spatial objects are usually tagged with semantic information like the type of an object (e.g., hospital, park), or the cuisine of a restaurant. In the DL world this information can be naturally represented in terms of concepts and roles.

Motivated by this, we present a rule-based framework and a toolbox to create benchmark instances from OSM datasets. Briefly, the main contributions are the following:

- We give a model-based formalization of OSM datasets which aims at abstracting from the currently employed but rather ad-hoc XML or object-relational representation. It allows one to view OSM maps as relational structures, possibly enriched with computable predicates like the spatial relations contains or next.
- Building on the above formalization, we present a rule-based language to extract information from OSM datasets (viewed as relational structures). In particular, a user can specify declarative rules which prescribe how to transform elements of an OSM map into ABox assertions. Different benchmark

---

[1] http://www.openstreetmap.org.
[2] E.g., visible in https://www.mapbox.com/osm-data-report/.

ABoxes can be created via small modifications of external functions, input parameter, and the rules of the specification.

- Our language is based on an extension of *Datalog*, which enjoys clear and well accepted semantics [1]. It has convenient features useful for benchmark generation.
- We have implemented a toolbox to create ABoxes from given input sources (e.g. an OSM database) and a given set of rules. The toolbox is highly configurable and can operate on various input/output sources, like RDF datasets, RDBMSs, and external functions. The input and result quality is measurable using descriptive statistics.
- By employing the above generation toolbox, we show on a proof-of-concept benchmark, how configurable and extensible the framework is. The toolbox has been already fruitfully been used for two benchmarks [5,7].

Our framework and toolbox provide an attractive means to develop tailored benchmarks for evaluating query answering systems, to gain new insights about them.

2 Formalization of OSM

In this section we formally describe our model for OSM data, which we later employ to describe our rule-based language to extract instance data from OSM data. Maps in OSM are represented using four basic constructs (a.k.a. *elements*):[3]

- *nodes*, which correspond to points with a geographic location;
- *geometries* (a.k.a. *ways*), which are given as sequences of nodes;
- *tuples* (a.k.a. *relations*), which are a sequences of nodes, geometries, and tuples;
- *tags*, which are used to describe metadata about nodes, geometries, and tuples.

Geometries are used in OSM to express polylines and polygons, in this way describing streets, rivers, parks, etc. OSM tuples are used to relate several elements, e.g. to indicate the turn priority in an intersection of two streets.

To formalize OSM maps, which in practice are encoded in XML, we assume infinite mutually disjoint sets $M_{nid}, M_{gid}, M_{tid}$ and M_{tags} of *node identifiers, geometry identifiers, tuple identifiers* and *tags*, respectively. We let $M_{id} = M_{nid} \cup M_{gid} \cup M_{tid}$ and call it the set of *identifiers*. An *(OSM) map* is a triple $\mathcal{M} = (\mathcal{D}, \mathcal{E}, \mathcal{L})$ as follows.

1. $\mathcal{D} \subseteq M_{id}$ is a finite set of identifiers called the *domain* of $\mathcal{M}$.
2. $\mathcal{E}$ is a function from $\mathcal{D}$ such that:
   (a) if $e \in M_{nid}$, then $\mathcal{E}(e) \in \mathbf{R} \times \mathbf{R}$;
   (b) if $e \in M_{gid}$, then $\mathcal{E}(e) = (e_1, \ldots, e_m)$ with $\{e_1, \ldots, e_m\} \subseteq \mathcal{D} \cap M_{nid}$;
   (c) if $e \in M_{tid}$, then $\mathcal{E}(e) = (e_1, \ldots, e_m)$ with $\{e_1, \ldots, e_m\} \subseteq \mathcal{D}$.
3. $\mathcal{L}$ is a *labeling* function $\mathcal{L} : \mathcal{D} \rightarrow 2^{M_{tags}}$.

---

[3] For clarity, we rename the expressions used in OSM.

Intuitively, in a map $\mathcal{M} = (\mathcal{D}, \mathcal{E}, \mathcal{L})$ the function $\mathcal{E}$ assigns to each node identifier a coordinate, to each geometry identifier a sequence of nodes, and to each tuple identifier a sequence of arbitrary identifiers.

Example 1. Assume we want to represent a bus route that, for the sake of simplicity, goes in a straight line from the point with coordinate $(0, 0)$ to the point with coordinate $(2, 0)$. In addition, the bus stops are at 3 locations with coordinates $(0, 0)$, $(1, 0)$ and $(2, 0)$. The names of the 3 stops are *S0*, *S1* and *S2*, respectively. This can be represented via the following map $\mathcal{M} = (\mathcal{D}, \mathcal{E}, \mathcal{L})$, where

- $\mathcal{D} = \{n_0, n_1, n_2, g, t\}$ with $\{n_0, n_1, n_2\} \subseteq \mathsf{M_{nid}}$, $g \in \mathsf{M_{gid}}$ and $t \in \mathsf{M_{tid}}$,
- $\mathcal{E}(n_0) = (0, 0)$, $\mathcal{E}(n_1) = (1, 0)$, $\mathcal{E}(n_2) = (2, 0)$,
- $\mathcal{E}(g) = (n_0, n_2)$ and $\mathcal{E}(t) = (g, n_0, n_1, n_2)$,
- $\mathcal{L}(n_0) = \{S0\}$, $\mathcal{L}(n_1) = \{S1\}$ and $\mathcal{L}(n_2) = \{S2\}$.

The tuple (g, n_0, n_1, n_2) encodes the 3 stops n_0, n_1, n_2 tied to the route given by g.

Enriching Maps with Computable Relations. The above formalizes the raw representation of OSM data. To make it accessible to rules, we allow to enrich maps with arbitrary computable relations over $\mathsf{M_{id}}$. In this way, we support incorporation of information that need not be given explicitly but can be computed from a map. Let $\mathsf{M_{rels}}$ be an infinite set of *map relation* symbols, each with an associated nonnegative integer, called the *arity*. An *enriched map* is a tuple $\mathcal{M} = (\mathcal{D}, \mathcal{E}, \mathcal{L}, \cdot^{\mathcal{M}})$, where $(\mathcal{D}, \mathcal{E}, \mathcal{L})$ is a map and $\cdot^{\mathcal{M}}$ is a partial function that assigns to a map relation symbol $R \in \mathsf{M_{rels}}$ a relation $R^{\mathcal{M}} \subseteq \mathcal{D}^n$, where n is the arity of R. In this way, a map can be enriched with externally computed spatial relations like the binary relations "is closer than 100m", "inside a country", "reachable from", etc. For the examples below, we assume that an enriched map $\mathcal{M}$ as above always defines the unary relation Tag_α for every tag $\alpha \in \mathsf{M_{tags}}$. In particular, we let $e \in \mathsf{Tag}_\alpha^{\mathcal{M}}$ iff $\alpha \in \mathcal{L}(e)$, where $e \in \mathcal{D}$. We will also use unary relations Point and Geom for points and geometries, and the binary relation Inside, where $\mathsf{Inside}(x, y)$ will mean that the point x is located inside the geometry y.

3   A Rule Language for Data Transformation

We define a rule-based language that can be used to describe how an ABox is created from an enriched map. Our language is based on *Datalog with stratified negation* [1].

Let $\mathsf{D_{rels}}$ be an infinite set of *Datalog relation symbols*, each with an associated *arity*. For simplicity, and with a slight abuse of notation, we assume that DL concept and role names form a subset of Datalog relations. Formally, we take an infinite set $\mathsf{D_{concepts}} \subseteq \mathsf{D_{rels}}$ of unary relations called *concept names* and an infinite set $\mathsf{D_{roles}} \subseteq \mathsf{D_{rels}}$ of binary relations called *role names*. Let $\mathsf{D_{vars}}$ be a countably infinite set of *variables*. Elements of $\mathsf{M_{id}} \cup \mathsf{D_{vars}}$ are called *terms*.

An *atom* is an expression $R(t)$ or *not* $R(t)$, where R is a map or a Datalog relation symbol of arity n, and t is an n-tuple of terms. We call $R(t)$ and *not* $R(t)$ a *positive atom* and a *negative atom*, respectively. A *rule* r is an expression of the form $B_1, \ldots, B_n \rightarrow H$, where $B_1, \ldots, B_n$ are atoms (called *body atoms*) and H is a positive atom with a Datalog relation symbol (called the *head atom*). We use $\mathsf{body}^+(r)$ and $\mathsf{body}^-(r)$ for the sets of positive and negative atoms in $\{B_1, \ldots, B_n\}$, respectively. We assume *(Datalog) safety*, i.e. each variable of r occurs in $\mathsf{body}^+(r)$. A *program* P is any finite set of rules. A rule or program is *ground* if it has no occurrences of variables. A rule r is *positive* if $\mathsf{body}^-(r) = \emptyset$. A program P is *positive* if all rules of P are positive. A program P is *stratified* if it can be partitioned into programs $P_1, \ldots, P_n$ such that:

(i) If $r \in P_i$ and *not* $R(t) \in \mathsf{body}^-(r)$, then there is no $j \geq i$ such that P_j has a rule with R occurring in the head.

(ii) If $r \in P_i$ and $R(t) \in \mathsf{body}^+(r)$, then there is no $j > i$ such that P_j has a rule with R occurring in the head.

The semantics of a program P is given relative to an enriched map $\mathcal{M}$. Its ground program $\mathsf{ground}(P, \mathcal{M})$ can be obtained from P by replacing in all possible ways the variables in rules of P with identifiers occurring in $\mathcal{M}$ or P. We use a variant of the Gelfond-Lifschitz reduct [9] to get rid of map atoms in a program. The *reduct* of P w.r.t. $\mathcal{M}$ is the program $P^{\mathcal{M}}$ obtained from $\mathsf{ground}(P, \mathcal{M})$ as follows:

(a) Delete from the body of every rule r every map atom *not* $R(t)$ with $t \notin R^{\mathcal{M}}$.

(b) Delete every rule r whose body contains a map atom *not* $R(t)$ with $t \in R^{\mathcal{M}}$.

Observe that $P^{\mathcal{M}}$ is an ordinary stratified Datalog program with identifiers acting as constants. We let $PM(\mathcal{M}, P)$ denote the *perfect model* of the program $P^{\mathcal{M}}$. See [1] for the construction of $PM(\mathcal{M}, P)$ by fix-point computation along the stratification. We are now ready to extract an ABox. Given a map $\mathcal{M}$ and a program P, we denote by $\mathsf{ABox}(\mathcal{M}, P)$ the restriction of $PM(\mathcal{M}, P)$ to the atoms over concept and role names.

We next illustrate some features of our rule language. The basic available service is to extract instances of concepts or roles by posing a standard conjunctive query over an OSM map. F.i., the following rule collects in the role `hasCinema` the cinemas of a city (we use sans-serif and typewriter font for map and Datalog relations, respectively):

$\mathsf{Point}(x), \mathsf{Tag}_{\mathsf{cinema}}(x), \mathsf{Geom}(y), \mathsf{Tag}_{\mathsf{city}}(y), \mathsf{Inside}(x, y) \rightarrow \mathtt{hasCinema}(y, x).$

Negation in rule bodies can be used for default, closed-world conclusions. E.g., the rule states that recreational areas include all parks that are not known to be private:

$\mathsf{Geom}(x), \mathsf{Tag}_{\mathsf{park}}(x), \textit{not}\ \mathsf{Tag}_{\mathsf{private}}(x) \rightarrow \mathtt{RecreationalArea}(x)$

Recursion is also useful and e.g., allows to deal with reachability, which appears naturally and in many forms in the context of geographic data.

E.g. suppose we want to collect pairs b_1, b_2 of bus stops such that b_2 is reachable from b_1 using public buses. To this end, we can assume the availability of an external binary relation hasStop which relates bus routes and their stops, i.e. hasStop(x, y) is true in case x is a geometry identifier corresponding to a bus route and y is a point identifier corresponding to a bus stop in the route represented by x. Then the desired pairs of bus stops can be collected in the role ReachByBus using the following recursive rules:

$$\text{hasStop}(x, y_1), \text{hasStop}(x, y_2) \rightarrow \text{ReachByBus}(y_1, y_2)$$

$$\text{ReachByBus}(y_1, y_2), \text{ReachByBus}(y_2, y_3) \rightarrow \text{ReachByBus}(y_1, y_3).$$

Extending the Rule Language with ETL Features. We introduce a custom language for the benchmark generation, which *extends* the Datalog language of the previous paragraph with *extract*, *transform*, and *load* (ETL) features. The combined language consists of *Data Source Declarations*, *Mapping Axioms*, and *Datalog Rules*. Data Source Declarations contain general definitions like RDBMS connection strings. A mapping axiom defines a single ETL step, where the syntax is an extension of the *Ontop* mapping language. It is defined either as a pair of *source* and *target* or as a triple of *source*, *transform*, and *target*: Each pair/triple has a first column containing a constant, a second column referring to the data source declarations, and a third column, which is modified depending on the source, target, or transformation line, respectively.[4]

4   Benchmarking Framework

The rule language $\mathcal{L}$ of the previous section gives us the means to define the data transformations. We combine the language with an OSM database $\mathcal{S}$, an input ontology $\mathcal{O}$ (a.k.a TBox), a set $\mathcal{Q}$ of conjunctive or SPARQL queries, the generation parameters $\mathcal{P}$, and external functions $\mathcal{E}$. The benchmark framework is denoted as $\mathcal{F} = \langle \mathcal{S}, \mathcal{O}, \mathcal{Q}, \mathcal{L}, \mathcal{P}, \mathcal{E} \rangle$ and produces a set of ABox instances denoted as $\mathcal{A} = (A_1, \ldots, A_n)$. Note that $\mathcal{F}$ might modify $\mathcal{O}$ slightly.

Workflow. The workflow of creating a benchmark and evaluating the respective reasoners can be split into an *initial* and a *repeating* part. The initial part consists of the following elements. First, one has to choose the ontology $\mathcal{O}$ and decide which ontology language should be investigated. The *ontology statistics* gives a first impression on the *expressivity* of the language such as DL-Lite$_R$ [4] or $\mathcal{EL}$ [2]. Then, $\mathcal{O}$ has to be customized (e.g., remove axioms) and loaded to the system. For $\mathcal{Q}$, either handcrafted queries (related to a practical domain) have to built or synthetic queries have to be generated (e.g., Sygenia [11]). After the initial part, we are able to generate the instance data for the fixed $\mathcal{O}$ and $\mathcal{Q}$. This part of the workflow can be repeated until certain properties are reached. It has the following steps:

---

[4] See the detailed syntax, prerequisites, and tools on https://github.com/ghxiao/city-bench.

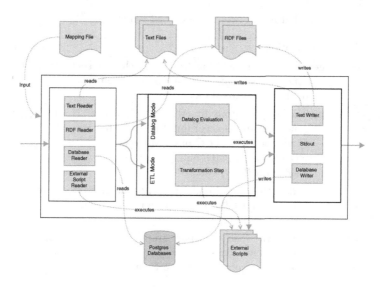

Fig. 1. System architecture, full lines are the control and dotted lines are the data flow

1. Creating an OSM database $\mathcal{S}$ with several instances, i.e., cities or countries;[4]
2. Applying *dataset statistics* to get a broad overview of the dataset, which leads to the selection of "interesting" datasets from $\mathcal{S}$;
3. Creating the rules of $\mathcal{L}$ to define the transformation for the instance generation and defining the parameters $\mathcal{P}$ and choosing the needed external functions of $\mathcal{F}$;
4. Calling the generation toolbox (see Sect. 5) and create the instances of $\mathcal{A}$;
5. Using *ABox statistics* to evaluate $\mathcal{A}$'s quality, if not satisfactory, repeat from 3.;

Descriptive Statistics. For the benchmark creation, descriptive statistics serves two purposes. First, it gives a broad picture of the datasets, which is important to formulate the mapping rules. Second, we use the statistics to guide and fine-tune the instance generation. That is, for generating the **next** relation, different distances can be calculated leading to different sizes of $\mathcal{A}$. Descriptive statistics can be applied on three levels. On the *ontology level*, ontology metrics regarding $\mathcal{O}$ can be produced using *owl-toolkit*[5] to calculate the number of concepts, roles, and axioms (e.g., sub-concept). On the *dataset level*, we provide general information on the selected OSM database instance including the main elements *Points*, *Lines*, *Roads*, and *Polygons* and details about *frequent item sets* [3] of keys and tags. On the *ABox level* we provide the statistics of the generated instances in $\mathcal{A}$. For this, we count the assertion in $\mathcal{A}$ for every atomic concept or role name of $\mathcal{O}$. For future work, we aim to estimate the instances based on the *subsumption graph* for the entire concept/role hierarchy.

---

[5] https://github.com/ghxiao/owl-toolkit.

Table 1. Available External Functions

Name	Description	Predicate
transformOSM	generates from OSM tags atoms which represent concepts/roles of $\mathcal{O}$. It has to be customized to the signature of $\mathcal{O}$.	$\mathsf{Tag}_{Park}(x)$
transformOSM-Random	instead of generating directly from OSM tags, it generates atoms according to a probabilities P assigned to a set O of OSM tags, e.g., $P(PublicPark)=0.8$ and $P(PrivatePark)=0.2$.	$\mathsf{Tag}_{P,O}(x)$
generateSpatial-Relation	generates the spatial relations **contains** or **next**, where a threshold parameter for the object distance can be given.	$\mathsf{next}_{10m}(x,y)$
generateStreet-Graph	generates the road/transport graph by creating instances for *edges* and *vertices* based on streets and corners between them.	$\mathsf{connected}(x,y)$ $\mathsf{Tag}_{corner}(x)$

External Functions and Parameters. External functions bridge the gap between $\mathcal{L}$ and external computations. They allow us to develop dataset-specific customization and functionalities, where the results (atoms) are associated with predicates of $\mathcal{L}$. In Table 1, we list the currently available external functions. In addition, we provide the functions *deleteRandom* and *deleteByFilter* which drop instances randomly or filter out instances from $\mathcal{A}$. The parameters are the means to fine-tune the generation. They are often not directly observable, hence we need the statistics tool to get a better understanding of data sources. From recent literature [16,20], we identified the following parameters for the instance generation:

- *ABox Size*: choice of the OSM instance (e.g., major cities or countries), but also by applying *deleteRandom* and *deleteByFilter*;
- *Degree of ABox Saturation:* can be indirectly manipulated by the use of Datalog rules in $\mathcal{L}$ to generate instances which otherwise would be deduced.
- *Distribution/Density of Nominal/Numeric Values:* input for *transformOSM-Random*;
- *Selectivity of Concept/Role Assertions:* input for *transformOSM* and *transformOSMRandom* and choice of OSM instances;
- *Graph Structure:* choice of OSM instance and selected graph (e.g., road vs. public transport network) for *generateStreetGraph*.

5   Implementation

We have developed for the framework a generation toolbox in Python 2.7, which is called as follows: `generate.py -m mapping.txt`

Modes. We provide two different modes with different evaluation strategies. The *Direct mode* is designed for simple bulk processing, where scalability and performance is crucial and complex calculations are moved to custom external scripts. We implemented the computation in a *data streaming*-based manner. The target components could be extended to custom triple stores like *Jena TDB*. The mapping axioms are evaluated in sequential order, hence dependencies between sources and targets are not considered. The *Datalog mode* extends the Direct mode and is designed for Datalog programs using the DLV system for evaluation.[6] The Datalog results are calculated in-memory and we follow a computation in three stages: 1. Previous ETL steps are evaluated to create the fact files for the EDB; 2. The defined Datalog programs (maintained in external files) are evaluated on the EDB files with the DLV module; 3. The (filtered) results (i.e., perfect models) are parsed and converted to tuples which then can be used by any target component. For now, we only handle a single model due to stratified Datalog.

Architecture. In Fig. 1, we show the architecture of the framework. It naturally results from the two modes and the source and target components. The following source and target components are implemented. For *Text files*, we use the standard functions of Python for reading, writing, and evaluating regular expressions. For *RDF files*, which are accessed by SPARQL queries, we leverage the functions of the *rdflib* library. At present for *RDBMSs*, we only include access to the spatial-extended RDBMS PostGIS 2.12 (for PostgreSQL), which is the most common system for OSM.

External Functions and Statistics. Besides the main script, we implemented Python scripts for the external functions from Sect. 4 for processing the (OSM) data. The list of implemented functions (e.g., `GenerateStreetGraph.py`) is available online.[4] `StatsOSM.py` and `StatsABox.py` are the statistical scripts for estimating the structure of the ABox and the main OSM elements. They calculate the values for the most used field/tag combinations. Additionally, we find the most frequent item sets using the *FP-Growth* algorithm.[7] For the ABox, we use the *rdflib* library to count and report the basic concept and role assertions.

6   Example Benchmark

In this section, we demonstrate how the framework can be applied to generate a proof-of-concept benchmark for OQA systems. All the mapping files, test datasets, and statistics are available online.[8]

OSM Dataset and Ontology. There are different subsets of different sizes and structures available for OSM. For this example, we chose the cities of *Cork, Riga, Bern, Vienna,* and *Berlin*.[9] Using the dataset statistics module for *Vienna,*

---

[6] http://www.dlvsystem.com/dlv/.

[7] https://github.com/enaeseth/python-fp-growth.

[8] https://github.com/ghxiao/city-bench/tree/master/benchmarks/rr2015.

[9] Downloaded on the 1.10.14 from http://download.bbbike.org/osm/bbbike/.

Table 2. Cities Dataset

City	$\#_{Points}$	$\#_{Lines}$	$\#_{Poly}$
Cork	6 068	14 378	4 934
Riga	19 172	43 042	67 708
Bern	68 831	83 351	151 195
Vienna	245 107	151 863	242 576
Berlin	236 114	218 664	430 652

Table 3. Road Network Instances

$\#_{Road}$	$\#_{Node}$	$\#_{connect}$	$\#_{Shop}$	$\#_{opr}$	$\#_{next_{50}}$
6 476	45 459	46 013	278	36	750
6 620	35 107	37 007	827	102	1 408
17 995	130 849	134 670	1 539	120	10 285
40 915	191 220	207 429	5 259	506	23 151
46 320	204 342	226 554	9 791	588	81 911

we observe for the field *Shop* 816 supermarkets, 453 hairdressers, 380 bakeries. For the field *Highway*, we have 29 392 residential, 4 087 secondary, and 3 973 primary streets.

The DL-Lite$_R$ [4] benchmark ontology is taken from the MyITS project [6]. It is tailored to geospatial and project specific data sources (e.g., a restaurant guide). The ontology is for OQA systems of average difficulty having only a few existential quantification on the right-hand side of the inclusion axioms. Due to its size and concept and role hierarchy depth, it poses a challenge regarding the rewritten query size.

Creation of a Road Network Benchmark. Besides creating the concept assertions for banks and shops, we extract the road network of the cities using the external function *generateStreetGraph* and encode the different roads into a single *road graph*. The road graph is represented by *nodes* which are asserted to the concept `Point` and by *edges* which are asserted to the role `connected`. By increasing the distances (e.g., from 50 m to 100 m) we saturate the `next` relation and generate more instances. Further, we use Datalog rules to calculate all paths (i.e. the transitive closure) of the street graph. The ABox statistics is shown in Tables 2 and 3; the cities are of increasing size, starting with *Cork* (25 000 objects) and ending with *Berlin* (885 000 objects).

7   Related Work

In addition to the "de facto" standard benchmark LUBM [10] and extended LUBM [16] with randomly generated instance data with a fixed ontology, several other works deal with testing OQA systems. They can be divided along conceptional reasoning, query generation, mere datasets, synthetic and real-life instance generation. The benchmarks provided by [18] consist of a set of ontologies and handcrafted queries, tailored for testing query rewriting techniques. These benchmarks are a popular choice for comparing the sizes of generated queries. In [20], the authors have provided tools to generate ABoxes for estimating the incompleteness of a given OQA system. In a similar spirit is [11], which provides tools to automatically generate conjunctive queries for testing correctness of OQA systems. The same authors also provide a collection of benchmarks for evaluating query rewriting systems [17], but they did not offer any novel generation tool. The work of [15] for OBDA is designed based on real data from the Norwegian Petroleum Directorate FactPages. However, it is focused solely on a fixed DL-Lite$_R$ ontology and queries. None of the above benchmarks provide large amounts of real-life instance data and an extended framework including various parameters and external functions. Furthermore, most of the mentioned approaches do not consider an iterative generation process using statistics to guide the generation. In the area of Spatial Semantic Web systems, a couple of benchmarks have been proposed to test geospatial extensions of SPARQL including the spatial extension of LUBM in [12] and the *Geographica* benchmark [8]. They are pre-computed and Queries geared towards testing spatial reasoning capabilities of systems, but not designed with OQA in mind.

8   Conclusion and Outlook

We have presented a flexible framework for generating instance data from a geospatial database for OQA systems. In particular, we have introduced a formalization of OSM and a Datalog-based mapping language as the formal underpinning of the framework. Datalog offers powerful features such as recursion and negation for benchmark generation. We have implemented an instance generation tool supporting the main *Datalog* mode and a simple *Direct* (extract-transform-load) mode for several types of input sources. Finally, we have demonstrated our approach on a proof-of-concept benchmark.

Future research is naturally directed to variants and extensions of the presented framework. We aim to extend the implementation to capture more input and output sources, further parameters (e.g. various degrees of graph connectedness) and services. Furthermore, a tighter integration of the Datalog solver/engine and the source/target components using dlvhex[10] is desired, which leads to a more efficient evaluation and more advanced capabilities (e.g., creating different ABoxes using all calculated answer sets). Another issue is to apply our framework to generate benchmarks for an extensive study of different OQA reasoners with different underlying technologies. Finally, the instance assertion statistics could be extended to the full subsumption graph.

References

1. Abiteboul, S., Hull, R., Vianu, V.: Foundations of Databases. Addison-Wesley, Boston (1995)
2. Baader, F., Brand, S., Lutz, C.: Pushing the $\mathcal{EL}$ envelope. In: Proceedings of IJCAI 2005, pp. 364–369. Morgan-Kaufmann Publishers (2005)
3. Borgelt, C.: Frequent item set mining. Data Min. Knowl. Disc. **2**(6), 437–456 (2012). Wiley Interdisciplinary Reviews
4. Calvanese, D., De Giacomo, G., Lembo, D., Lenzerini, M., Rosati, R.: Tractable reasoning and efficient query answering in description logics: the dl-lite family. J. Autom. Reasoning **39**(3), 385–429 (2007)
5. Eiter, T., Fink, M., Stepanova, D.: Computing repairs for inconsistent DL-programs over $\mathcal{EL}$ ontologies. In: Fermé, E., Leite, J. (eds.) JELIA 2014. LNCS, vol. 8761, pp. 426–441. Springer, Heidelberg (2014)

---

[10] http://www.kr.tuwien.ac.at/research/systems/dlvhex/.

6. Eiter, T., Krennwallner, T., Schneider, P.: Lightweight spatial conjunctive query answering using keywords. In: Cimiano, P., Corcho, O., Presutti, V., Hollink, L., Rudolph, S. (eds.) ESWC 2013. LNCS, vol. 7882, pp. 243–258. Springer, Heidelberg (2013)

7. Eiter, T., Schneider, P., Simkus, M., Xiao, G.: Using openstreetmap data to create benchmarks for description logic reasoners. In: Informal proceedings of ORE 2014, July 2014

8. Garbis, G., Kyzirakos, K., Koubarakis, M.: Geographica: a benchmark for geospatial rdf stores (Long Version). In: Alani, H., Kagal, L., Fokoue, A., Groth, P., Biemann, C., Parreira, J.X., Aroyo, L., Noy, N., Welty, C., Janowicz, K. (eds.) ISWC 2013, Part II. LNCS, vol. 8219, pp. 343–359. Springer, Heidelberg (2013)

9. Gelfond, M., Lifschitz, V.: The stable model semantics for logic programming. In: Proceedings of ICLP/SLP 1988, vol. 88, pp. 1070–1080 (1988)

10. Guo, Y., Pan, Z., Heflin, J.: LUBM: a benchmark for OWL knowledge base systems. Web Semantics 3(2–3), 158–182 (2005)

11. Imprialou, M., Stoilos, G., Cuenca Grau, B., Benchmarking ontology-based query rewriting systems. In: Proceedings of AAAI 2012 (2012)

12. Kolas, D: A benchmark for spatial semantic web systems. In: 4th International Workshop on Scalable Semantic Web Knowledge Base Systems (SSWS2008), October 2008

13. Kollia, I., Glimm, B.: Optimizing SPARQL query answering over OWL ontologies. J. Artif. Intell. Res. (JAIR) 48, 253–303 (2013)

14. Kontchakov, R., Rezk, M., Rodríguez-Muro, M., Xiao, G., Zakharyaschev, M.: Answering SPARQL queries over databases under OWL 2 QL entailment regime. In: Mika, P., Tudorache, T., Bernstein, A., Welty, C., Knoblock, C., Vrandečić, D., Groth, P., Noy, N., Janowicz, K., Goble, C. (eds.) ISWC 2014, Part I. LNCS, vol. 8796, pp. 552–567. Springer, Heidelberg (2014)

15. Lanti, D., Rezk, M., Xiao, G., Calvanese, D.: The NPD benchmark: reality check for OBDA systems. In: Proceedings of EDBT 2015. ACM Press (2015)

16. Ma, L., Yang, Y., Qiu, Z., Xie, G.T., Pan, Y., Liu, S.: Towards a complete OWL ontology benchmark. In: Sure, Y., Domingue, J. (eds.) ESWC 2006. LNCS, vol. 4011, pp. 125–139. Springer, Heidelberg (2006)

17. Mora, J., Corcho, O.: Towards a systematic benchmarking of ontology-based query rewriting systems. In: Alani, H., Kagal, L., Fokoue, A., Groth, P., Biemann, C., Parreira, J.X., Aroyo, L., Noy, N., Welty, C., Janowicz, K. (eds.) ISWC 2013, Part II. LNCS, vol. 8219, pp. 376–391. Springer, Heidelberg (2013)

18. Pérez-Urbina, H., Horrocks, I., Motik, B.: Efficient query answering for OWL 2. In: Bernstein, A., Karger, D.R., Heath, T., Feigenbaum, L., Maynard, D., Motta, E., Thirunarayan, K. (eds.) ISWC 2009. LNCS, vol. 5823, pp. 489–504. Springer, Heidelberg (2009)

19. Sirin, E., Parsia, P., Cuenca Grau, B., Kalyanpur, A., Katz, Y.: Pellet: a practical OWL-DL reasoner. J. Web Sem. 5(2), 51–53 (2007)

20. Stoilos, G., Cuenca Grau, B., Horrocks, I.: How incomplete is your semantic web reasoner? In: Proceedings of AAAI 2010. AAAI Press (2010)

Web Stream Reasoning in Practice:
On the Expressivity vs. Scalability Tradeoff

Stefano Germano[1][⊠], Thu-Le Pham[2], and Alessandra Mileo[2]

[1] Department of Mathematics and Computer Science,
University of Calabria, Rende, Cosenza, Italy
germano@mat.unical.it
[2] Insight Centre for Data Analytics, National University of Ireland,
Galway, Ireland
{thule.pham,alessandra.mileo}@insight-centre.org

Abstract. Advances in the Internet of Things and the Web of Data created huge opportunities for developing applications that can generate actionable knowledge out of streaming data. The trade-off between scalability and expressivity is a key challenge in this setting, and more investigation is required to identify what are the relevant features in optimizing this trade-off, and what role do they have in the optimization. In this paper we motivate the need for heuristics to design adaptive solutions and, following an empirical approach, we highlight some key concepts and ideas that can guide the design of heuristics for adaptive optimization of Web Stream Reasoning.

Keywords: Stream reasoning · Logic programming · Semantic Web

1 Introduction and Background

Web Stream Reasoning has emerged as a research field that explores advances in Semantic Web technologies for representing and processing data streams on one hand, and emerging approaches to perform complex rule-based inference over dynamic and changing environments on the other hand. Advances in the Internet and Sensor technologies converging to the Internet of Things (IoT) have also contributed to the creation of a plethora of new applications that require to process and make sense of web data streams in a scalable way.

In the Semantic Web and Linked Data realm, technologies such as RDF, OWL, SPARQL have been recently extended to provide mechanisms for processing semantic data streams [1–3]. However a variety of real-world applications in the IoT space require reasoning capabilities that can handle incomplete, diverse and unreliable input and extract actionable knowledge from it. Non-monotonic stream reasoning techniques for the (Semantic) Web have potential impact on tackling them.

Semantic technologies for handling data streams can not exhibit complex reasoning capabilities such as the ability of managing defaults, common-sense, preferences, recursion, and non-determinism. Conversely, logic-based non-monotonic

© Springer International Publishing Switzerland 2015
B. ten Cate and A. Mileo (Eds.): RR 2015, LNCS 9209, pp. 105–112, 2015.
DOI: 10.1007/978-3-319-22002-4_9

reasoners can perform such tasks but are suitable for data that changes in low volumes at low frequency.

To reach the goal of combining the advantages of these two approaches in the last years a few works have been proposed; some tried to develop extensions of ASP [4] in order to deal with dynamic data [5–7], others tried to combine semantic stream query processing and non-monotonic reasoning [8,9]. The StreamRule framework [9] is an example which provides a baseline for exploring the applicability of complex reasoning on Semantic Web Streams.

The conceptual idea behind StreamRule is to process data streams at different levels of abstraction and granularity, in such a way to guarantee that the amount of relevant data is filtered (and therefore reduced in size) as the complexity of the reasoning increases.[1] This has in principle a high potential in making complex reasoning on semantic streams feasible and scalable. However, the one-directional processing pipeline in StreamRule from query evaluation to non-monotonic reasoning is a strong limitation in exploring the expressivity vs. scalability trade-off: the dynamic nature of web streams and their changing rate, quality and relevance makes it impossible to specify at design time what is the correct throughput and reasoning complexity the system can support and what window size and time-decay model is most suitable.

The main goal of this paper is to provide a preliminary analysis on how we can improve the scalability of expressive stream reasoning for the Semantic Web combining continuous query processing and Answer Set Programming (ASP). We rely on the StreamRule [9] system as an instance of such an approach for implementation and testing, and we aim at providing general insights that holds for any ASP-based stream reasoning system.

The main idea we present in this paper relies on concepts that can help make the StreamRule processing pipeline bi-directional or adaptive, so that the expressivity vs. scalability trade-off can be optimized in changing environments.

We start our investigation by identifying in Sect. 2 which the key features can potentially affect the expressivity vs. scalability trade-off in a 2-tier web stream reasoning system like StreamRule. The correlation between such features and their impact on scalability are empirically evaluated in Sect. 3 by our practical analysis of performance and correlation between streaming rate, window size, properties of the input streams and complexity of the reasoning. Some hints for discussion are presented based on our empirical results.

2   Core Concepts for Analysis

The key contribution of this position paper is to report on initial investigation on how to perform complex reasoning on web data streams maintaining scalability. We refer to scalability as to the ability to provide answers in an acceptable time with increasing input size and when the reasoning gets computationally intensive.

---

[1] Note that in ASP, the expressivity of the language is strictly related to the computational complexity, therefore we refer to expressivity and (computational) complexity interchangeably throughout the paper.

We will introduce some key concepts that can later guide the design of heuristics for systems like StreamRule (which we will consider as a reference model in the remainder of this paper), where query processing and non-monotonic reasoning features are adapted to continuously improve the expressivity versus scalability trade-off in changing environments. The conceptual architecture of StreamRule is based on a 2-tier approach to web stream reasoning where query processing (first tier) is used to filter semantic data elements, while non-monotonic reasoning (second tier) is used for computationally intensive tasks as shown in Fig. 1.

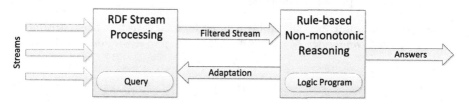

Fig. 1. 2-tier approach

We define the following concepts and notation:

Unit of Time (U). The unit of time is the time interval to which collected inputs are sent to the system (we will assume this as fixed in our analysis).

Reasoning complexity (C). We refer to the reasoning complexity as the computational complexity required to perform a given reasoning task involving a set of ASP rules. As mentioned earlier in this paper, the computational complexity is strictly related to the language expressivity in ASP; in fact, more expressive language constructs in ASP correspond to higher computational complexity. The type of rules used within the ASP program affects grounding (which also affects memory consumption) and solving (which is related to computational complexity), and therefore has an impact on scalability. For simplicity we assume in our analysis that the reasoning complexity is fixed (based on the rules in the program). However, this aspect deserves a more formal characterization to be able to used the reasoning complexity as a feature to design adaptive heuristics for optimization, and we plan to investigate this in future work.

Streaming size (S). The streaming size is the number of input elements sent to the reasoning component every Unit of Time.

Window size (W). The (tuple based) window size[2] is the size of the input the reasoning component processes per computation.

Reasoning time (R_t). The reasoning time is considered as the time needed by the nonmonotonic reasoner (second tier only) to compute a solution.[3]

---

[2] In this paper we only consider non-overlapping windows. For overlapping windows, the formula $T_\omega(S, W)$ should hold also when duplicating events in overlapping parts.

[3] Note that this is different from the total processing time, which includes the time required for query processing (first tier). In this paper, we mainly focus on the reasoning time only, relying on the extensive evaluation of query processing engines for the query processing time [1].

$T(N)$ is the time needed by the reasoner to process N input elements.

$T_\omega(S, W)$ is the time needed by the reasoner to process a streaming size S dividing (and processing) it into windows of size W. The number of windows (and therefore the number of computations needed) is $\lceil \frac{S}{W} \rceil$. Formally $T_\omega(S, W) = \lceil \frac{S}{W} \rceil \times T(W)$.

S_u is the number of elements that can be processed by the reasoner in one unit of time.[4] Formally $S_u = N$ s.t. $T(N) = 1$.

S_l is the maximum number of elements that can be processed within one unit of time using a proper windows size W.

The question summarizing our problem is as follows: *Given a fixed streaming size S with fixed complexity C and unit of time U, find a window size W such that*

$$T_\omega(S, W) \leq U \tag{Q1}$$

Finding this window size and being able to adapt it to changing streaming rates would reduce the bottleneck between the two tiers, since it will ensure that the nonmonotonic reasoner can keep up with the results produced by the query processing engine without the cumulative delay experienced in StreamRule. Previous experiments in [9] showed that the current implementation of StreamRule with CQELS [1] as query processor and Clingo [5] as ASP reasoner encounters a bottleneck when the non-monotonic reasoner returns results after the next input arrives from the stream query processing component, thus cumulating a delay that makes the system not scalable. Making the process bi-directional requires to dynamically provide answers to Q1.

We can observe that if $T(N)$ is monotonically increasing, we have that

$$\forall S' \ where \ S_u \leq S' \leq S_l, \ \exists W' \ s.t. \ T_\omega(S', W') \leq U$$

This is our case, as illustrated in our empirical evaluation.

3   Experiments and Discussion

In this section, we are presenting the scenario, dataset, ruleset, and platform we used for the empirical evaluation of our trade-off analysis and discuss our findings.

Scenario. Consider a user moving on a path. She wants to know real-time events that affect her travel plan to react accordingly. The stream reasoning system receives events as Linked Data Streams that indicate changes in the real world (such as accidents, road traffic, flooding, road diversions and so on) and updates on the user's current status (such as user's location and activity).

---

[4] Note that this is different from the streaming size.

With this information as input stream, the Web Stream Reasoning system is in charge of (i) selecting among the list of events, which are the ones that are really relevant according to the user's context, and (ii) continuously ranking their level of criticality with respect to the user task and context[5], in order to decide whether a new path needs to be computed.

The query processing component filters out events which are unrelated to the user, e.g. events are not on the user's path, thus limiting the input size for the nonmonotonic reasoner. The reasoner receives as input filtered events and an instance of the context ontology related to the activities and status of the user, provides ranked critical events as output. The event includes 4 attributes: type, value, time, and location. For example:

event(weather, strong-wind, 2014-11-26T13:00:00, 38011736-121867224)

describes the condition of weather being strong wind at a certain time and a given location.

Dataset. For our experiment, we generate traffic events based on 10 types of events such as: roadwork, obstructions, incident, sporting events, disasters, weather, traffic conditions, device status, visibility air quality, incident response status. Each type of events has more than 2 values, e.g. traffic condition has values: good, slow, congested. In addition to that we create a small instance of the context ontology which describes the effect of events on certain activities.

Ruleset. The ASP rule set we used for this experiment includes 10 rules which have 2 negated atoms (using negation-as-failure). Since the complexity of the reasoning is fixed and related to a specific program in our setup, we do not quantify the complexity in this initial investigation.

Platform. We used the state-of-the-art ASP reasoner Clingo 4.3.0 and Java 1.7. The experiment were conducted over a machine running Debian GNU/Linux 6.0.10, containing 8-cores of 2.13 GHz processor and 64 GB RAM.

Empirical Results. We evaluated the same ASP program with varying input size S (from 100 to 30000 events) and measured the reasoning time of the system $(T(S))$. We trigger the reasoner 20 times for each S and then we computed the *interquartile mean (IQM)* to smooth results. These values are plotted in Fig. 2.

Given $U = 1\,s$, the graph shows $S_u = 17520$ *events*. In other words, for this particular case (and fixed Ruleset and Platform), the stream reasoning system will be "stable" if the streaming size of the ASP reasoner is smaller than 17520 *events*. For streaming size bigger than 17520 *events*, the system will cumulate a delay that will cause a bottleneck. Giving our function is monotonically increasing, there are some streaming sizes bigger than S_u that can be processed in less

---

[5] In the current implementation we evaluate criticality mainly based on how close an event is to the user location, and how fast is the user moving. In future work we plan to extend this contextual characterization to consider not only location but also other features such as the user transportation type, user's health condition etc.

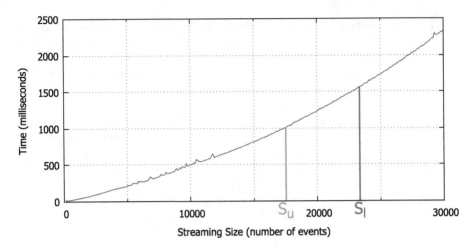

Fig. 2. Reasoning time

than 1 U. We then investigated the idea of dividing events in windows, assuming we can find a split such that the correctness of the result won't be affected[6].

The easiest way to perform this split is to consider several windows of the same size. For example, consider $S = 20000$ *events*, it will take $1232\,ms$ for the reasoner process all 20000 *events* in one computation ($T(20000) = 1232\,ms$). The whole system will combine a delay in each computation and therefore will crash at some point. However, if we use the window size $W = 5000$ *events* ($T(5000) = 216\,ms$), the reasoner will take $T_\omega(20000, 5000) = \lceil \frac{20000}{5000} \rceil \times T(5000) = 4 \times 216\,ms = 864\,ms$ for processing S, using 4 computations. So we have found a proper window size (W) such that $T_\omega(S, W) \leq U$; in other words we have found a split for which the system remain stable.

Moreover, if we divide S into windows of size $W = 2000$ *events*, the reasoning time for S will be $T_\omega(20000, 2000) = 720\,ms$, so also in this case T_ω is less than or equal to 1 U. Therefore, in general, there could be more than one way to split the events.

For any given S, a proper value[7] for W such that $T_\omega(S, W) \leq U$ can be found in a trivial way just checking for each streaming size (S') less than S_u the time required ($T(S')$) and verifying that $T_\omega(S, S') = \lceil \frac{S}{S'} \rceil \times T(S') \leq U$; when we find such S', we can put $W = S'$.

Running this algorithm increasing S up to the point where we cannot find any streaming size S' less than S_u such that $T_\omega(S, S') \leq U$, we can compute the value of S_l. Based on this we found that for this experiment the value of S_l is 23350 *events*. It means that the system can scale if the streaming size is less than or equal to 23350 *events*.

---

[6] Algorithms to perform such splits are under investigation and will be the subject of future work.

[7] Note that our goal is not to find the minimum, we just want to find one split.

We can also apply these algorithms to the trend line function that fit the data in order to have a more precise result. In our experiment we found an Order 2 polynomial trend line which fit very well our data with an R-squared value of 0.99979 and we have used this to find the values of S_u and S_l.

Discussion and Conclusions. Based on our experiments, we observe that:

– Given a unit of time and a particular ASP program, we can find an optimal window size for a given streaming size for reducing the processing time of the system.
– This conclusion holds if there is no dependency between input events for the reasoning component[8].

We are currently investigating how to generalize our empirical results to set the basis for designing adaptive heuristics for Web Stream Reasoning. An improvement to this approach, for some scenarios, is to find the value of W that minimize $T_\omega(S, W)$. A key aspect we are also considering is to provide a formal characterization that helps relaxing the assumption of independence between input events, in order to determine how to find an optimal number of window for a given streaming rate and a given ASP program. Since we started our empirical evaluation based on a given ASP program, another interesting directions will be to investigate more in-depth how the complexity of the reasoning affects our analysis.

Acknowledgments. This research has been partially supported by Science Foundation Ireland (SFI) under grant No. SFI/12/RC/2289 and EU FP7 CityPulse Project under grant No.603095. http://www.ict-citypulse.eu

References

1. Le-Phuoc, D., Dao-Tran, M., Xavier Parreira, J., Hauswirth, M.: A native and adaptive approach for unified processing of linked streams and linked data. In: Aroyo, L., Welty, C., Alani, H., Taylor, J., Bernstein, A., Kagal, L., Noy, N., Blomqvist, E. (eds.) ISWC 2011, Part I. LNCS, vol. 7031, pp. 370–388. Springer, Heidelberg (2011)
2. Barbieri, D.F., Braga, D., Ceri, S., Valle, E.D., Grossniklaus, M.: Querying RDF streams with C-SPARQL. ACM SIGMOD Rec. **39**(1), 20–26 (2010)
3. Calbimonte, J.-P., Corcho, O., Gray, A.J.G.: Enabling ontology-based access to streaming data sources. In: Patel-Schneider, P.F., Pan, Y., Hitzler, P., Mika, P., Zhang, L., Pan, J.Z., Horrocks, I., Glimm, B. (eds.) ISWC 2010, Part I. LNCS, vol. 6496, pp. 96–111. Springer, Heidelberg (2010)
4. Baral, C.: Knowledge representation, reasoning and declarative problem solving. Cambridge University Press, Cambridge (2003)

---

[8] Note that this assumption needs to be formally characterized and more investigation is ongoing in this direction.

5. Gebser, M., Grote, T., Kaminski, R., Obermeier, P., Sabuncu, O., Schaub, T.: Stream reasoning with answer set programming: preliminary report. In: KR, vol. 12, pp. 613–617 (2012)
6. Gebser, M., Grote, T., Kaminski, R., Obermeier, P., Sabuncu, O., Schaub, T.: Answer set programming for stream reasoning. CoRR (2013)
7. Beck, H., Dao-Tran, M., Eiter, T., Fink, M.: Lars: A logic-based framework for analyzing reasoning over streams. In: AAAI (2015)
8. Do, T.M., Loke, S.W., Liu, F.: Answer set programming for stream reasoning. In: Butz, C., Lingras, P. (eds.) Canadian AI 2011. LNCS, vol. 6657, pp. 104–109. Springer, Heidelberg (2011)
9. Mileo, A., Abdelrahman, A., Policarpio, S., Hauswirth, M.: StreamRule: a non-monotonic stream reasoning system for the semantic web. In: Faber, W., Lembo, D. (eds.) RR 2013. LNCS, vol. 7994, pp. 247–252. Springer, Heidelberg (2013)

A Procedure for an Event-Condition-Transaction Language

Ana Sofia Gomes$^{(\boxtimes)}$ and José Júlio Alferes

NOVA-LINCS - Departamento de Informática, Faculdade Ciências e Tecnologias
Universidade NOVA de Lisboa, Lisboa, Portugal
sofia.gomes@campus.fct.unl.pt

Abstract. Event-Condition-Action languages are the commonly accepted paradigm to express and model the behavior of reactive systems. While numerous Event-Condition-Action languages have been proposed in the literature, differing e.g. on the expressivity of the language and on its operational behavior, existing Event-Condition-Action languages do not generally support the action component to be formulated as a transaction. In this paper, sustaining that it is important to execute transactions in reactive languages, we propose an Event-Condition-Transaction language, based on an extension of Transaction Logic. This extension, called Transaction Logic with Events ($\mathcal{TR}^{ev}$), combines reasoning about the execution of transactions with the ability to detect complex events. An important characteristic of $\mathcal{TR}^{ev}$ is that it takes a choice function as a parameter of the theory, leaving open the behavioral decisions of the logic, and thereby allowing it to be suitable for a wide-spectrum of application scenarios like Semantic Web, multi-agent systems, databases, etc. We start by showing how $\mathcal{TR}^{ev}$ can be used as an Event-Condition-Action language where actions are considered as transactions, and how to differently instantiate this choice function to achieve different operational behaviors. Then, based on a particular operational instantiation of the logic, we present a procedure that is sound and complete w.r.t. the semantics and that is able to execute $\mathcal{TR}^{ev}$ programs.

1 Introduction and Motivation

Most of today's applications are highly dynamic, as they produce, or depend on data that is rapidly changing and evolving over time. This is especially the case of Web-based applications, which normally have to deal with large volumes of data, with new information being added and updated constantly. In fact, with the amount of online data increasing exponentially every year, it is estimated that the annual IP traffic will reach the zettabyte threshold in 2015[1].

This sheer amount of information forced us to dramatically change the way we store, access, and reason with data on the web, and led to the development

---

A.S. Gomes and J.J. Alferes—This work was supported by project ERRO (PTDC/EIA-CCO/121823/2010).

[1] http://blogs.cisco.com/news/the-dawn-of-the-zettabyte-era-infographic

© Springer International Publishing Switzerland 2015
B. ten Cate and A. Mileo (Eds.): RR 2015, LNCS 9209, pp. 113–129, 2015.
DOI: 10.1007/978-3-319-22002-4_10

of several research areas. Among them, the Semantic Web, which started about 15 years ago, aims to enrich the Web with machine-understandable information by promoting a collaborative movement where users publish data in one of the standard formats, RDF or OWL, designed to give a precise semantic meaning to web data. On the other hand, research areas like Event Processing (EP) and Stream Reasoning deal with the problem of handling and processing a high volume of events (also called a *stream*), and reason with them to detect meaningful event patterns (also known as complex events). Although EP started in the 1990 s within the database community, today we can find a number of EP solutions based on Semantic Web technologies like RDF, SPARQL and OWL [3,17,20]. These EP solutions deal with the challenge of detecting event patterns on a stream of atomic events. But detecting these patterns is only part of what one has to do to deal with the dynamics of data. In fact, detecting event patterns is only meaningful if we can act upon the knowledge of their occurrence.

Event-Condition-Action (ECA) languages solve this by explicitly defining what should be the reaction of a system, when a given event pattern is detected. For that, ECA-rules have the standard form: *on* **event** *if* **condition** *do* **action**, where whenever an event is known to be true, the condition is checked to hold in the current state and, if that is the case, the action is executed. Initially introduced to support reactivity in database systems, numerous ECA languages have been proposed e.g. in the context of the Semantic Web [6,10,19], multi-agent systems [12,15,18], conflict resolution [11]. Moreover, although all these languages share the same reactive paradigm, they often vary on the expressivity of the language and the connectives available, but also on their operational behavior, namely on the event consumption details, and on the response policy and scheduling. While ECA languages started in the database context, and many solutions exist supporting rich languages for defining complex actions, most ECA languages do not allow the action component to be defined as a transaction, or when they do, they either lack from a declarative semantics (e.g. [19]), or can only be applied in a database context since they only detect atomic events defined as primitive insertions/deletes on the database (e.g. [16,24]).

Originally proposed to make database operations reliable, transactions ensure several properties, like consistency or atomicity, over the execution of a set of actions. We sustain that in several situations ECA languages are indeed required to execute transactions in response to events, where either the whole transaction is executed or, if anything fails meanwhile, nothing is changed in the knowledge base (KB). As an application scenario, consider the case where the local government wants to control the city's air quality by restricting vehicles more than 15 years old to enter certain city areas. For that, the city needs to identify the plates of the cars crossing these areas, and issue fines for the unauthorized vehicles. In this case, whenever a vehicle enters the controled area (the event), we need to check if that vehicle has access to the area (the condition), and if not, issue a fine and notify the driver for the infraction. Clearly, some transactional properties regarding these actions must be ensured, as it can never be the case that a fine is issued and the driver is not notified, or vice-versa.

Transaction Logic ($\mathcal{TR}$) is a general logic proposed in [8] to deal with transactions, by allowing us to reason about their behaviors but also to *execute* them. For that, $\mathcal{TR}$ provides a general model theory that is parametric on a pair of oracles defining the semantics of states and updates of the KB (e.g. relational databases, action languages, description logics). With it, one can reason about the sequence of states (denoted as paths) where a transaction is executed, *independently* of the semantics of states and primitive actions of the KB. Additionally, $\mathcal{TR}$ also provides a proof-theory to *execute* a subclass of $\mathcal{TR}$ programs that can be formulated as Horn-like clauses. However, $\mathcal{TR}$ cannot simultaneously deal with complex events and transactions, and for that we have previously proposed $\mathcal{TR}^{ev}$ in [14]. $\mathcal{TR}^{ev}$ is an extension of $\mathcal{TR}$ that, like the original $\mathcal{TR}$, can reason about the execution of transactions, but also allows for the definition of complex events by combining atomic (or other complex) events. In $\mathcal{TR}^{ev}$, atomic events can either be external events, which are signalled to the KB, or the execution of primitive updates in the KB (similarly e.g. to the events "on insert" in databases). Moreover, as in active databases, transactions in $\mathcal{TR}^{ev}$ are *constrained* by the events that occur during their execution, as a transaction can only successfully commit when all events triggered during its execution are addressed. Importantly, $\mathcal{TR}^{ev}$ is parameterized with a pair of oracles as in the original $\mathcal{TR}$, but also with a *choice* function abstracting the semantics of a reactive language from its response policies decisions. $\mathcal{TR}^{ev}$ is a first step to achieve a reactive language that combines the detection of events with the execution of transactions. However, it leaves open how the ECA paradigm can be encoded and achieved in $\mathcal{TR}^{ev}$, but also, how can one execute such reactive ECA rules.

In this paper we show how one can indeed use $\mathcal{TR}^{ev}$ as the basis for an Event-Condition-Transaction language, illustrating how different response policies can be achieved by correctly instantiating the choice functions. Then, to make the logic useful in practice, we provide a proof procedure sound and complete with the semantics, for a Horn-like subset of the logic, and for a particular choice function instantiation.

2   Background: $\mathcal{TR}^{ev}$

Transaction Logic [8], $\mathcal{TR}$, is a logic to execute and reason about general changes in KB, when these changes need to follow a transactional behavior. In a nutshell[2], The $\mathcal{TR}$ syntax extends that of first order logic with the operators $\otimes$ and $\Diamond$, where $\phi \otimes \psi$ denotes the action composed by an execution of ϕ followed by an execution of ψ, and $\Diamond\phi$ denotes the hypothetical execution of ϕ, i.e. a test to see whether ϕ can be executed but leaving the current state unchanged. Then, $\phi \wedge \psi$ denotes the simultaneous execution of ϕ and ψ; $\phi \wedge \psi$ the execution of ϕ or ψ; and $\neg\phi$ an execution where ϕ is not executed.

In $\mathcal{TR}$ all formulas are read as transactions which are evaluated over sequences of KB states known as *paths*, and satisfaction of formulas means execution.

---

[2] For lack of space, and since $\mathcal{TR}^{ev}$ is an extension of $\mathcal{TR}$ (cf. [14]) we do not make a thorough overview of $\mathcal{TR}$ here. For complete details see e.g. [8,14].

I.e., a formula (or transaction) ϕ is true over a path π iff the transaction success-fully executes over that sequence of states. $\mathcal{TR}$ makes no particular assumption on the representation of states, or on how states change. For that, $\mathcal{TR}$ requires the existence of two oracles: the data oracle $\mathcal{O}^d$ abstracting the representation of KB states and used to query them, and the transition oracle $\mathcal{O}^t$ abstracting the way states change. For example, a KB made of a relational database [8] can be modeled by having states represented as sets of ground atomic formulas, where the data oracle simply returns all these formulas, i.e., $\mathcal{O}^d(D) = D$. Moreover, for each predicate p in the KB, the transition oracle defines p.ins and p.del, repre-senting the insertion and deletion of p, respectively (where p.ins $\in \mathcal{O}^t(D_1, D_2)$ iff $D_2 = D_1 \cup \{\text{p}\}$ and, p.del $\in \mathcal{O}^t(D_1, D_2)$ iff $D_2 = D_1 \setminus \{\text{p}\}$). We will use this relational database oracle definition in our example (Example 1).

The logic provides the concept of a *model* of a $\mathcal{TR}$ theory, which allows one to prove properties of transactions that hold for every possible path of execution; and the notion of executional entailment, in which a transaction ϕ is entailed by a theory given an initial state D_0 and a program P (denoted as $P, D_0 - \models \phi$), if there is a path $D_0, D_1 \ldots, D_n$ starting in D_0 on which the transaction, as a whole, succeeds. Given a transaction and an initial state, the executional entailment provides a means to determine what should be the evolution of states of the KB, to succeed the transaction in an atomic way. Non-deterministic transactions are possible, and in this case several successful paths exist. Finally, a proof procedure and corresponding implementation exist for a special class of $\mathcal{TR}$ theories, called serial-Horn programs, which extend definite logic programs with serial conjunction.

$\mathcal{TR}^{ev}$ extends $\mathcal{TR}$ in that, besides dealing with the execution of transaction, it is also able to raise and detect complex events. For that, $\mathcal{TR}^{ev}$ separates the evaluation of events from the evaluation of transactions. This is reflected in its syntax, and on the two different satisfaction relations – the event satisfaction $\models_{ev}$ and the transaction satisfaction $\models$. The alphabet of $\mathcal{TR}^{ev}$ contains an infinite number of constants $\mathcal{C}$, function symbols $\mathcal{F}$, variables $\mathcal{V}$ and predicate symbols $\mathcal{P}$. Furthermore, predicates in $\mathcal{TR}^{ev}$ are partitioned into transaction names ($\mathcal{P}_t$), event names ($\mathcal{P}_e$), and oracle primitives ($\mathcal{P}_\mathcal{O}$). Finally, formulas are also partitioned into transaction formulas and event formulas.

Event formulas are formulas meant to be *detected* and are either an event occurrence, or an expression defined inductively as $\neg\phi$, $\phi \wedge \psi$, $\phi \vee \psi$, or $\phi \otimes \psi$, where ϕ and ψ are event formulas. We further assume $\phi; \psi$, which is syntactic sugar for $\phi \otimes \text{path} \otimes \psi$ (where path is just any tautology, cf. [9]), with the meaning: "ϕ and then ψ, but where arbitrary events may be true between ϕ and ψ". An *event occurrence* is of the form $\mathbf{o}(\varphi)$ s.t. $\varphi \in \mathcal{P}_e$ or $\varphi \in \mathcal{P}_\mathcal{O}$ (the latter are events signaling changes in the KB, needed to allow reactive rules similar to e.g. "on insert" triggers in databases).

Transaction formulas are formulas that can be *executed*, and are either a transaction atom, or an expression defined inductively as $\neg\phi$, $\Diamond\phi$, $\phi \wedge \psi$, $\phi \vee \psi$, or $\phi \otimes \psi$. A *transaction atom* is either a transaction name (in $\mathcal{P}_t$), an oracle defined primitive (in $\mathcal{P}_\mathcal{O}$), the response to an event (written $\mathbf{r}(\varphi)$ where $\varphi \in \mathcal{P}_\mathcal{O} \cup \mathcal{P}_e$), or an event name (in $\mathcal{P}_e$). The latter corresponds to the (trans)action of *explicitly* triggering/raising an event directly in a transaction. Finally, rules have the form $\varphi \leftarrow \psi$ and can be transaction or (complex) event rules. In a transaction rule φ is a transaction atom and ψ a transaction formula; in an event rule φ is an event occurrence and ψ is an event formula. A *program* is a set of transaction and event rules.

Example 1. Consider the example from the introduction, and a relational KB with information about license plates registration, drivers addresses, and fines. The transaction of processing an unauthorized access of a vehicle V, at a given time T, and city area A (written processUnAccess(V, T, A)) can be defined in $\mathcal{TR}^{ev}$ (and in $\mathcal{TR}$) as:[3]

processUnAccess(V, T, A) ←
    fineCost(A, Cost) ⊗ unAccess(V, T, A, Cost).ins ⊗ notifyFine(V, T, A, Cost)
notifyFine(V, T, A, Cost) ←
    isDriver(D, V) ⊗ hasAddress(D, Addr) ⊗ sendLetter(D, A, V, T, A, Cost)

One can also define complex events. E.g., a vehicle is said to enter the city from a given entrance E_1, if it is detected by the two sensors of that entrance, first by sensor 1 and then by sensor 2, with a time difference less than 0.5s:

o(enterCityE$_1$(V, T$_2$)) ←
    ((o(sensor1E$_1$(V, T$_1$).ins); o(sensor2E$_2$(V, T$_2$).ins)) ∧ T$_2$ − T$_1$ < 0.5)

For simplicity, in this example we assume that sensors' data is directly inserted in the application's database, where the semantics of $\_ins$ is as defined by the relational oracle above. However, other paradigms can be used to represent and reason about sensor's data. Namely, and given some recent works in the field of Sensor Networks [21, 22], we could have alternatively assumed a Semantic Sensor Network to publish data in RDF, and have defined the oracle $\mathcal{O}^d$ to query this data using SPARQL. This would allow us to use $\mathcal{O}^d$ to integrate RDF data with the government's database.

Central to the theory of $\mathcal{TR}^{ev}$ is the correspondence between $\mathbf{o}(\varphi)$ and $\mathbf{r}(\varphi)$. As a transactional system, the occurrence of an event constrains the satisfaction path of the transaction where the event occurs, and a transaction can only "commit" if all the occurring events are answered. More precisely, a transaction is only satisfied on a path, if all the events occurring over that path are properly responded to. This behavior is achieved by evaluating event occurrences and transactions differently, and by imposing $\mathbf{r}(\varphi)$ to be true in the paths where $\mathbf{o}(\varphi)$ holds. For dealing with cases where more than one occurrence holds simultaneously, $\mathcal{TR}^{ev}$ takes as parameter, besides $\mathcal{TR}$'s data and transition oracles, a *choice* function selecting what event should be responded at a given time, in case of conflict. This function abstracts the operational decisions from the logic, and allows $\mathcal{TR}^{ev}$ to be useful for a wide spectrum of applications.

---

[3] Without loss of generality (cf. [9]), we consider Herbrand instantiations of the language.

As in $\mathcal{TR}$, formulas of $\mathcal{TR}^{ev}$ are evaluated over paths (sequence of states), and the theory allows us to reason about *how* does the KB evolve in a transactional way, based on an initial KB state. In addition, a path in $\mathcal{TR}^{ev}$ is of the form $D_0 \overset{O_1}{\to} \ldots \overset{O_n}{\to} D_n$, and where each O_i is a primitive event occurrence that holds in the state transition D_{i-1}, D_i. As a reactive system, $\mathcal{TR}^{ev}$ receives a series (or a stream) of external events which may cause the execution of transactions in response.

This is defined as $P, D_0 - \models e_1 \otimes \ldots \otimes e_k$, where D_0 is the initial KB state and $e_1 \otimes \ldots \otimes e_k$ is the sequence of external events that arrived. Then, we want to know what is the path $D_0 \overset{O_1}{\to} \ldots \overset{O_n}{\to} D_n$ encoding a KB evolution that responds to $e_1 \otimes \ldots \otimes e_k$.

As usual, satisfaction of formulas is based on interpretations which define what atoms are true over what paths, by mapping every possible path to a set of atoms. If a transaction (resp. event) atom ϕ belongs to $M(\pi)$ then ϕ is said to execute (resp. occur) over path π given interpretation M. However, only the mappings compliant with the specified oracles are considered as interpretations:

Definition 1 (Interpretation). *An interpretation is a mapping M assigning a set of atoms (or $\top^4$) to paths, with the restrictions (where D_is are states, and φ an atom):*

1. $\varphi \in M(\langle D \rangle)$ if $\varphi \in \mathcal{O}^d(D)$
2. $\{\varphi, \mathbf{o}(\varphi)\} \subseteq M(\langle D_1 \overset{\mathbf{o}(\varphi)}{\to} D_2 \rangle)$ if $\varphi \in \mathcal{O}^t(D_1, D_2)$
3. $\mathbf{o}(e) \in M(\langle D \overset{\mathbf{o}(e)}{\to} D \rangle)$

Satisfaction of formulas requires the definition of operations on paths. E.g., $\phi \otimes \psi$ is true on a path if ϕ is true up to some point in the path, and ψ is true from that onwards.

Definition 2 (Path Splits, Subpaths and Prefixes). *Let π be a k-path, i.e. a path of length k of the form $\langle D_1 \overset{O_1}{\to} \ldots \overset{O_{k-1}}{\to} D_k \rangle$. A split of π is any pair of subpaths, π_1 and π_2, s.t. $\pi_1 = \langle D_1 \overset{O_1}{\to} \ldots \overset{O_{i-1}}{\to} D_i \rangle$ and $\pi_2 = \langle D_i \overset{O_i}{\to} \ldots \overset{O_{k-1}}{\to} D_k \rangle$ for some i ($1 \leq i \leq k$). In this case, we write $\pi = \pi_1 \circ \pi_2$. A subpath π' of π is any subset of states of π where both the order of the states and their annotations is preserved. A prefix π_1 of π is any subpath of π sharing the initial state.*

Satisfaction of complex formulas is different for event formulas and transaction formulas. While the satisfaction of event formulas concerns the *detection* of an event, the satisfaction of transaction formulas concerns the *execution* of actions in a transactional way. As such, when compared to the original $\mathcal{TR}$, transactions in $\mathcal{TR}^{ev}$ are further required to execute *all* the responses of the events occurring in the original execution path of that transaction. In other words, a transaction φ is satisfied over a path π, if φ is executed on a prefix π_1 of π (where $\pi = \pi_1 \circ \pi_2$), and all events occurring over π_1 are *responded*

---

[4] For not having to consider partial mappings, besides formulas, interpretations can also return the special symbol $\top$. The interested reader is referred to [8] for details.

over π_2. This requires a non-monotonic behavior of the satisfaction relation of transaction formulas, making them dependent on the satisfaction of events.

Definition 3 (Satisfaction of Event Formulas). *Let M be an interpretation, π a path and ϕ a formula. If $M(\pi) = \top$ then $M, \pi \models_{ev} \phi$; else:*

1. **Base Case:** $M, \pi \models_{ev} \phi$ *iff $\phi \in M(\pi)$ for every event occurrence ϕ*
2. **Negation:** $M, \pi \models_{ev} \neg\phi$ *iff it is not the case that $M, \pi \models_{ev} \phi$*
3. **Disjunction:** $M, \pi \models_{ev} \phi \vee \psi$ *iff $M, \pi \models_{ev} \phi$ or $M, \pi \models_{ev} \psi$.*
4. **Serial Conjunction:** $M, \pi \models_{ev} \phi\otimes\psi$ *iff there is a split $\pi_1 \circ \pi_2$ of π s.t. $M, \pi_1 \models_{ev} \phi$ and $M, \pi_2 \models_{ev} \psi$*
5. **Executional Possibility:** $M, \pi \models_{ev} \Diamond\phi$ *iff π is a 1-path of the form $\langle D \rangle$ for some state D and $M, \pi' \models_{ev} \phi$ for some path π' that begins at D.*

Definition 4 (Satisfaction of Transaction Formulas). *Let M be an interpretation, π a path, ϕ transaction formula. If $M(\pi) = \top$ then $M, \pi \models \phi$; else:*

1. **Base Case:** $M, \pi \models p$ *iff there is a prefix π' of π s.t. $p \in M(\pi')$ and π is an expansion of path π' w.r.t. M, for every transaction atom p s.t. $p \notin \mathcal{P}_e$.*
2. **Event Case:** $M, \pi \models e$ *iff $e \in \mathcal{P}_e$ and there is a prefix π' of π s.t. $M, \pi' \models_{ev} \mathbf{o}(e)$ and π is an expansion of path π' w.r.t. M.*
3. **Negation:** $M, \pi \models \neg\phi$ *iff it is not the case that $M, \pi \models \phi$*
4. **Disjunction:** $M, \pi \models \phi \vee \psi$ *iff $M, \pi \models \phi$ or $M, \pi \models \psi$.*
5. **Serial Conjunction:** $M, \pi \models \phi \otimes \psi$ *iff there is a prefix π' of π and a split $\pi_1 \circ \pi_2$ of π' s.t. $M, \pi_1 \models \phi$ and $M, \pi_2 \models \psi$ and π is an expansion of path π' w.r.t. M.*
6. **Executional Possibility:** $M, \pi \models \Diamond\phi$ *iff π is a 1-path of the form $\langle D \rangle$ for some state D and $M, \pi' \models \phi$ for some path π' that begins at D.*

The latter definition depends on the notion of expansion of a path. An *expansion* of a path π_1 w.r.t. to an interpretation M is an operation that returns a new path π_2 where all events occurring over π_1 (and also over π_2) are completely answered. Formalizing this expansion requires the prior definition of what it means to answer an event:

Definition 5 (Path response). *For a path π_1 and an interpretation M we say that π is a response of π_1 iff $choice(M, \pi_1) = e$ and we can split π into $\pi_1 \circ \pi_2$ s.t. $M, \pi_2 \models \mathbf{r}(e)$.*

The choice of what unanswered event should be picked at each moment is given by an event function *choice*. This function has the role to decide what events are unanswered over a path π w.r.t. an interpretation M and, based on a given criteria, select what event among them should be responded to first. Just like $\mathcal{TR}$ is parametric to a pair of oracles ($\mathcal{O}^d$ and $\mathcal{O}^t$), $\mathcal{TR}^{ev}$ takes the *choice* function as an additional parameter. For now, we leave the definition and role of this function open until Sect. 3. Nevertheless, and importantly, if all events that occur on a path π are answered on π w.r.t. M, then $choice(M, \pi) = \epsilon$. We can now define what is an expansion of a path.

Definition 6 (Expansion of a path). *A path π is completely answered w.r.t. to an interpretation M iff $choice(M, \pi) = \epsilon$. π is an expansion of the path π_1 w.r.t. M iff:*

- π *is completely answered w.r.t.* M, *and*
- *either* $\pi = \pi_1$; *or there is a sequence of paths* $\pi_1, \ldots, \pi$, *starting in* π_1 *and ending in* π, *s.t. each* π_i *in the sequence is a response of* π_{i-1} *w.r.t.* M.

The latter definition specifies how to expand a path π_1 in order to obtain another path π where all events satisfied over subpaths of π are also answered within π. This must perforce have some procedural nature: it must start by detecting which are the unanswered events; pick one of them, according to some criteria given by a *choice* function, that for now is seen as a parameter; then expand the path with the response of the chosen event. Each path π_i of the sequence $\pi_1, \pi_2, \ldots, \pi$ is a prefix of the path π_{i+1}, and where at least one of the events unanswered on π_i is now answered on π'; otherwise, if all events occurring over π_i are answered, then $\pi_i = \pi$, and the expansion is complete. We can now define the notion of *model* of formulas and programs.

Definition 7 (Models and Minimal Models). *An interpretation* M *is a model of a transaction (resp. event) formula* ϕ *iff for every path* π, $M, \pi \models \phi$ *(resp.* $M, \pi \models_{ev} \phi$). M *is a model of a program* P *(denoted* $M \models P$) *iff it is a model of every rule in* P. *We say that a model is minimal if it is a* $\subseteq$-*minimal model.*

This notion of models can be used to reason about properties of transaction and event formulas that hold for *every* possible path of execution. However, to know if a formula succeeds on a particular path, we need only to consider the event occurrences *supported* by that path, either because they appear as occurrences in the transition of states, or because they are a necessary consequence of the program's rules given that path. Because of this, executional entailment in $\mathcal{TR}^{ev}$ is defined w.r.t. minimal models.

Definition 8 ($\mathcal{TR}^{ev}$ Executional Entailment). *Let* P *be a program,* ϕ *a transaction formula and* $D_1 \xrightarrow{O_0} \ldots \xrightarrow{O_n} D_n$ *a path. Then* $P, (D_1 \xrightarrow{O_0} \ldots \xrightarrow{O_n} D_n) \models \phi$ *($\star$) iff for every minimal model* M *of* P, $M, \langle D_1 \xrightarrow{O_0} \ldots \xrightarrow{O_n} D_n \rangle \models \phi$. $P, D_1 - \models \phi$ *is said to be true, if there is a path* $D_1 \xrightarrow{O_0} \ldots \xrightarrow{O_n} D_n$ *that makes ($\star$) true.*

3 $\mathcal{TR}^{ev}$ as an Event-Condition-Transaction and the *choice* function

In the previous section, we provided the logical background for a reactive language that can express and reason about both transactions and complex events. Next we show how $\mathcal{TR}^{ev}$ can indeed be used as an Event-Condition-Transaction language, and how several ECA behaviors can be embedded in the semantics by providing the right translation into $\mathcal{TR}^{ev}$, and the right instantiations of the *choice* function.

As mentioned, an ECA language follows the basic paradigm **on** *event* **if** *condition* **do** *action*, defining that, whenever the *event* is learned to occur, the *condition* is tested, and, if it holds, the *action* is executed. As such, an ECA rule is said to be active whenever the event holds and, to satisfy it, either the condition does not hold before the execution of the action, or the action is issued for execution. Moreover, in an Event-Condition-Transaction language, this action needs not only to be executed, but to be executed as a transaction. This means that we need to guarantee that either the whole of the transaction is executed or, if anything fails meanwhile, the KB is left unchanged.

Since $\mathcal{TR}^{ev}$ forces every formula $\mathbf{r}(ev)$ to be true whenever $\mathbf{o}(ev)$ is learnt to be true, this behavior can be simply encoded as:

$$\mathbf{r}(ev) \leftarrow \Diamond cond \otimes action$$
$$\mathbf{r}(ev) \leftarrow \neg\Diamond cond \tag{1}$$

where $\Diamond cond$ is a test (and which necessarily does not cause changes in the KB) to determine if the condition $cond$ holds, and if this is the case, the $action$ is executed in the KB. Moreover, if one wants to define the event ev as a complex event, then one should add a rule stating the event pattern definition: $\mathbf{o}(ev) \leftarrow body$.

Example 2. Recall the example in the introduction and the transaction defined in Example 1 for processing unauthorized accesses. Then event-condition-transaction rule triggering that transaction can be written as:

$$\mathbf{o}(\texttt{enterCity}(V, T, e1)) \leftarrow \mathbf{o}(\texttt{enterCityE}_1(V, T))$$
$$\mathbf{r}(\texttt{enterCity}(V, \_, A)) \leftarrow \texttt{authorized}(V, A)$$
$$\mathbf{r}(\texttt{enterCity}(V, T, A)) \leftarrow \texttt{unauthorized}(V, A) \otimes \texttt{processUnAccess}(V, T, A)$$

where, since `authorized` and `unauthorized` are queries to the KB (i.e., they cause no change in the database), we can drop the $\Diamond$ constructor.

Moreover, the definition of an ECA language requires the specification of an operational behaviors, which in turn, involves two majors decisions: 1) in which order should events be responded when more than one event is detected simultaneously; and 2) how should an event be responded to. In order to make $\mathcal{TR}^{ev}$ as flexible as possible, its model theory was abstracted from these decisions, encapsulating them in a *choice* function. This function is required as a parameter of the theory (similarly to the oracles $\mathcal{O}^t$ and $\mathcal{O}^d$) and precisely defines what is the next event that still needs to be responded.

Definition 9 (*choice* **function**). *Let M be an interpretation and π be a path. Then:*
$$choice(M, \pi) = firstUnans(M, \pi, order(M, \pi)).$$

Matching these two major decisions, our definition of the *choice* function is partitioned in two functions: the *order* function specifying the sorting criteria of events, and a *firstUnans* function which checks what events are unanswered and returns the first one based on the previous order. The former decision defines the handling order of events, i.e. given a set of occurring events, what should be responded first. This ordering can be defined e.g. based on when they have occurred (temporal order), on a priority list, or any other criteria. This decision defines the response policy of an ECA-language, i.e. how should an event be responded. We start by illustrating the *order* function:

Example 3 (Ordering-Functions). Let $\langle e_1, \ldots, e_n \rangle$ be a sequence of events, π a path, and M an interpretation.

Temporal Ending Order $order(M, \pi) = \langle e_1, \ldots, e_n \rangle$ iff $\forall e_i$ s.t. $1 \leq i \leq n$ then $\exists \pi_i$ subpath of π where $M, \pi_i \models_{ev} \mathbf{o}(e_i)$ and $\forall e_j$ s.t. $i < j$ and then e_j occurs after e_i w.r.t. π.

Temporal Starting Order $order(M, \pi) = \langle e_1, \ldots, e_n \rangle$ iff $\forall e_i$ s.t. $1 \leq i \leq n$ then $\exists \pi_i$ subpath of π where $M, \pi_i \models_{ev} \mathbf{o}(e_i)$ and $\forall e_j$ s.t. $i < j$ then e_j starts before e_i w.r.t. π.

Priority List Order Let L be a priority list where events are linked with numbers starting in 1, where 1 is the event with higher priority. $order_L(M, \pi) = \langle e_1, \ldots, e_n \rangle$ iff $\forall e_i \ \exists \pi_i$ subpath of π s.t. $M, \pi_i \models_{ev} \mathbf{o}(e_i)$ and $\forall e_j$ where $1 \leq i < j \leq n$, π_j is subpath of π and $M, \pi_j \models_{ev} \mathbf{o}(e_j)$ then $L(e_i) \leq L(e_j)$.

All these examples require the notion of event ordering, which can be defined as:

Definition 10 (Ordering of Events). *Let e_1, e_2 be events, π a path, and M an interpretation. e_2 occurs after e_1 w.r.t. π and M iff $\exists \pi_1, \pi_2$ subpaths of π s.t. $\pi_1 = \langle D_i \stackrel{O_i}{\longrightarrow} \ldots \stackrel{O_{j-1}}{\longrightarrow} D_j \rangle$, $\pi_2 = \langle D_n \stackrel{O_n}{\longrightarrow} \ldots \stackrel{O_{m-1}}{\longrightarrow} D_m \rangle$, $M, \pi_1 \models_{ev} \mathbf{o}(e_1)$, $M, \pi_2 \models_{ev} \mathbf{o}(e_2)$ and $D_j \leq D_m$ w.r.t. the ordering in π. e_1 starts before e_2 w.r.t. π if $D_i \leq D_n$*

Choosing the appropriate event ordering obviously depends on the application in mind. For instance, in system monitoring applications there may exist alarms with higher priority over others that need to be addressed immediately, while in a webstore context it may be more important to treat events in the temporal orders in which they are detected.

It remains to be defined the response policy, i.e., what requisites should be imposed w.r.t. the response executions. This is done by appropriately instantiating the *firstUnans* function. Here we illustrate two alternative instantiations. In the first, encoded in *Relaxed Response*, the function simply retrieves the first event e such that its response is not satisfied in a path after the occurrence. With it, if an event occurs more than once, it is sufficient to respond to it once. Alternative definitions are possible, e.g. where responses are issued explicitly for each event. This is encoded in the *Explicit Response*, where we verify whether $\mathbf{r}(e_i)$ is satisfied but always w.r.t. its correct order.

Example 4 (Answering Choices). Let π be a path, M an interpretation, and $\langle e_1, \ldots, e_n \rangle$ a sequence of events.

Relaxed Response $firstUnans(M, \pi, \langle e_1, \ldots, e_n \rangle) = e_i$ if e_i is the first event in $\langle e_1, \ldots, e_n \rangle$ s.t. $\exists \pi'$ subpath of π where $M, \pi' \models_{ev} \mathbf{o}(e)$ and $\neg \exists \pi''$ s.t. π'' is also a subpath of π, π'' is after π' and $M, \pi'' \models \mathbf{r}(e)$.

Explicit Response $firstUnans(M, \pi, \langle e_1, \ldots, e_n \rangle) = e_i$ if e_i is the first event in $\langle e_1, \ldots, e_n \rangle$ s.t. $\exists \pi'$ subpath of π where $M, \pi' \models_{ev} \mathbf{o}(e)$ and if $\exists \pi''$ subpath of π that is after π' where $M, \pi'' \models \mathbf{r}(e_i)$ then $\exists \pi_1, \pi_2$ subpaths of π and π_2 is after π_1 where $M, \pi_1 \models_{ev} \mathbf{o}(e_j)$, $M, \pi_2 \models \mathbf{r}(e_j)$, $j < i$ and π'' starts before the ending of π_2

4 Procedure for Serial-$\mathcal{TR}^{ev}$

To make $\mathcal{TR}^{ev}$ useful in practice, we propose a proof procedure for executing Event-Condition-Transaction rules that is sound and complete w.r.t. $\mathcal{TR}^{ev}$'s executional entailment. Given a $\mathcal{TR}^{ev}$ program and a KB state, the procedure takes a *stream* of events that are known to occur, and finds a KB evolution where all the (possibly complex) events resulting from the direct and indirect occurrence of this stream, are responded to as a transaction. More precisely, the procedure finds solutions for statements of the form $P, D_0 - \models e_1 \otimes \ldots \otimes e_k$, by finding paths (which encode a KB evolution) where the formula $P, D_0 \overset{O_0}{\rightarrow} \ldots \overset{O_{n-1}}{\longmapsto} D_n \models e_1 \otimes \ldots \otimes e_k$ holds. Regarding the (procedural) choices discussed in the previous section, this procedures fixes an event ordering based on a priority list, and assumes Explicit Response (cf. Example 4).

The procedure is partitioned into two major parts: the execution of actions (based on a top-down computation), and the detection of event patterns (based on a bottom-up computation). The detection of event patterns is inspired by the ETALIS detection algorithm [4], where event rules are first pre-processed using a *binarization of events*. In other words, event rules are first transformed so that all their bodies have at most two atoms. If we have a body with more than 2 atoms, e.g. $\mathbf{o}(e) \leftarrow \mathbf{o}(a)$ OP $\mathbf{o}(b)$ OP $\mathbf{o}(c)$ and if we assume a left-associative operator, the binarization of this rule leads to replacing it by $ie_1 \leftarrow \mathbf{o}(a)$ OP $\mathbf{o}(b)$ and $\mathbf{o}(e) \leftarrow ie_1$ OP $\mathbf{o}(c)$, and where OP can be any binary $\mathcal{TR}^{ev}$ operator. This is done without loss of generality since:

Proposition 1 (Program equivalence). *Let P_1 and P_2 be programs, π be a path and ϕ a formula defined in both P_1 and P_2 alphabet. We say that $P_1 \equiv P_2$ if:*

$$M_1, \pi \models \phi \quad \textit{iff} \quad M_2, \pi \models \phi$$

where M_1 (resp. M_2) is the set of minimal models of P_1 (resp. P_2).

Let P be a program with rule: $e \leftarrow e_1$ OP e_2 OP e_3 for any events e_1-e_3 and operator OP, and let P' be obtained from P by removing that rule and adding $ie_1 \leftarrow e_1$ OP e_2 and $e \leftarrow ie_1$ OP e_3. Then $P \equiv P'$.

Besides binarization, and as it is usual in EP systems, we restrict the use of negation in the procedure. The problem with negation is that it is hard to detect the non-presence of an event pattern in an unbounded interval. Due to this, EP systems like [1,4] always define negation bounded to two other events as in $\mathbf{not}(e_3)[e_1, e_2]$. This holds if e_3 does not occur in the interval defined by the occurrence of e_1 and e_2. This is captured in $\mathcal{TR}^{ev}$ by: $e_1 \otimes \neg(\mathbf{path} \otimes e_3 \otimes \mathbf{path}) \otimes e_2$, and we restrict negation to this pattern.

The execution of actions is based on $\mathcal{TR}$ proof-theory [8,13] which is only defined for a subclass of programs where transactions can be expressed as serial-Horn goals. As such, the execution of transactions in this procedure is defined for this fragment, which resembles Horn-clauses of logic programming. A serial goal is a transaction formula of the form $a_1 \otimes a_2 \otimes \ldots \otimes a_n$ where each a_i is an atom and $n \geq 0$. When $n = 0$ we write $()$ which denotes the empty goal. Finally, a serial-Horn rule has the form $b \leftarrow a_1 \otimes \ldots \otimes a_n$, where the body $a_1 \otimes \ldots \otimes a_n$ is a serial goal and b is an atom.

The procedure starts with a program P, an initial state D, a serial goal $e_1 \otimes \ldots \otimes e_k$ and iteratively manipulates *resolvents*. At each step, the procedure non-deterministically applies a series of rules of Definition 11 to the current resolvent until it either reaches the empty goal and succeeds, or no more rules are applicable and the derivation fails. Moreover, if the procedure succeeds, it also returns a path in which the goal succeeds. To cater for this last requirement, resolvents contain the path obtained so far. A resolvent is of the form $\pi, ESet \Vdash_{P'}^{id} \phi$, where ϕ is the current transaction goal to be executed, π is the path obtained by the procedure so far, $ESet$ is the set of events that were previously triggered and still need to be addressed, and id is an auxiliar state count identifier (whose usage is made clear below). Finally, P' is the current program, containing all the rules from the original program P plus temporary event rules to help deal with the detection of event patterns. A successful derivation for $P, D- \models e_1 \otimes \ldots \otimes e_k$ starts with the resolvent $\langle D \rangle, \emptyset \Vdash_P^1 e_1 \otimes \ldots \otimes e_k$ and ends in the resolvent $\pi, \emptyset \Vdash_{P'}^{id} ()$. If such a derivation exists, then we write $P, \pi \vdash e_1 \otimes \ldots \otimes e_k$. Most derivation rules have a direct correspondence with $\mathcal{TR}$'s proof theory [8], but now incorporating the notion of path expansion and event detection. Rule 1 replaces a transaction atom L by the *Body* of a program rule whose head is L; rule 2 deals with a query to the oracle deleting it from the set of goals whenever this query is true in the current state (i.e., the last state of π); rule 3 executes actions according to the transition oracle definition (i.e., if an action A can be executed in the last state D_1 of path π, reaching the state D_2, then we add the path $D_1 \xrightarrow{\mathbf{o}(A)} D_2$ to our current path, but also the path *expansion*, cf. Definition 12, resulting from answering the events that have directly or indirectly become true because of $\mathbf{o}(A)$); finally, rule 4 deals with the triggering of explicit events – if an event e is explicitly triggered and D_1 is the last state of the current path π, then we add the information that $\mathbf{o}(e)$ occurred in state D_1 to our path, and expand it with the KB evolution needed to answer all events that are now true because of e's occurrence.

Definition 11 (Execution). *A derivation for a serial goal ϕ in a program P and state D is a sequence of resolvents starting with $\langle D \rangle, \emptyset \Vdash_P^1 \phi$, and obtained by non-deterministically applying the following rules.*

Let $\pi, ESet \Vdash_{P_1}^{id} L_1 \otimes L_2 \otimes \ldots \otimes L_k$ be a resolvent. The next resolvent is:

1. **Unfolding of Rule:**
   $\pi, ESet \Vdash_{P_1}^{id} Body \otimes L_2 \otimes \ldots \otimes L_k$ *if* $L_1 \leftarrow Body \in P$
2. **Query:**
   $\pi, ESet \Vdash_{P_1}^{id} L_2 \otimes \ldots \otimes L_k$ *if* $\mathtt{last}(\pi) = D_1$ *and* $\mathcal{O}^d(D_1) \models L_1$
3. **Update primitive:**
   $\pi \circ \langle D_1 \xrightarrow{\mathbf{o}(L_1)} D_2 \xrightarrow{O_2} \ldots \xrightarrow{O_{j-1}} D_j \rangle, ESet' \Vdash_{P_2}^{id'} L_2 \otimes \ldots \otimes L_k$ *if:*
   - $\mathtt{last}(\pi) = D_1$ *and* $\mathcal{O}^t(D_1, D_2) \models L_1$
   - $ExpandPath(P_1, L_1, \pi, ESet, id) = (P_2, \langle D_2 \xrightarrow{O_2} \ldots \xrightarrow{O_{j-1}} D_j \rangle, ESet', id')$
4. **Explicit event request:**
   $\pi \circ \langle D_1 \xrightarrow{\mathbf{o}(L_1)} D_1 \xrightarrow{O_1} \ldots \xrightarrow{O_{j-1}} D_j \rangle, ESet' \Vdash_{P_2}^{id'} L_2 \otimes \ldots \otimes L_k$ *if:*

- $\texttt{last}(\pi) = D_1$ and $L_1 \in \mathcal{P}_e$
- $ExpandPath(P_1, L_1, \pi, ESet, id) = (P_2, \pi \circ \langle D_1 \xrightarrow{O_1} \dots \xrightarrow{O_{j-1}} D_j \rangle, ESet', id')$

An execution for ϕ in program P, and state D is successful *if it ends in a resolvent of the form* $\pi, \emptyset \Vdash_{P'}^{id'} ()$. *In this case we write* $P, \pi \vdash \phi$.

The *ExpandPath* function is called in the procedure above whenever an event A (either an explicit event or a the execution of a primitive action) occurs, and is responsible for expanding the current path with the responses of the events made true. Moreover, given the event rules in the program, it may be the case that other event occurrences become true, and these other events need also to be responded by this function. The set of events that become true due to the occurrence of A are computed by the *Closure* function (Definition 14), and the process is iterated until there are no more unanswered events:

Definition 12 (Expand Path).
Input: program P, primitive A, path π, event set $ESet$, id
Output: program P', path π', event set $ESet'$, id'
Define $ESet' := ESet \cup \{o(A)\{id, id+1\}\}$, $\pi' := \pi$, $id' := id + 1$, $P' := P$
while $\texttt{needResponse}(ESet', id, id') \neq \emptyset$ {

1. *Let $(ESet_{temp}, P_{temp}) = Closure(ESet', P', id')$*
2. *Let $o(e) = FirstInOrder(\texttt{needResponse}(ESet_{temp}, id, id'))$*
3. *Let D be the final state of π', and consider a derivation starting in $\langle D \rangle, \emptyset \Vdash_{P_{temp}}^{id'} r(e)$*
   and ending in $\pi_f, ESet_f \Vdash_{P_f}^{id_f} ()$; if no such derivation exists **return** *failure.*
4. *Define $\pi' := \pi' \circ \pi_f$, $id' := id_f$, $P' := P_f$, $ESet' := (ESet_{temp} \cup ESet_f) \setminus \{o(e)\}$ }*

If $\texttt{needResponse}(ESet', id, id') = \emptyset$ *then the computation is said to be* successful.
In this case **return** $(P', \pi', ESet', id')$; *otherwise* **return** *failure.*

The *ESet'* variable in the latter definition contains, at each moment, the set of events that have happened during the execution. Each event in this set is associated to a pair $\{id_i, id_f\}$ specifying the exact interval where the event happened. One can then recast the path of occurrence based on these *ids*. If $e\{id_i, id_f\} \in ESet$, and $\pi = \langle D_1 \xrightarrow{O_1} \dots \xrightarrow{O_{k-1}} D_k \rangle$ is the current path, then event e is said to occur in $\pi_{<id_i, id_f>}$, where $\pi_{<id_i, id_f>}$ is the path obtained from π by trimming it from state D_{id_i} to state D_{id_f}: $\langle D_{id_i} \xrightarrow{O_{id_i}} \dots \xrightarrow{O_{id_f-1}} D_{id_f} \rangle$.

At each iteration *ExpandPath* collects all the events in *ESet'* which need to be responded to w.r.t. that iteration. I.e., the events whose occurrence holds on a path starting after the initial state of the function call:

Definition 13 ($\texttt{needResponse}(ESet, id_i, id_j)$). *Let id_1, id_2, id_i be identifiers and $ESet$ a set of events of the form $e\{id_1, id_2\}$, where id_1 and id_2 define the starting and ending of e, respectively. $\texttt{needResponse}(ESet, id_i, id_j)$ is the subset of $ESet$ s.t. $id_i \leq id_1 \leq id_2 \leq id_j$.*

FirstInOrder function simply sorts the events w.r.t. the chosen order function, according to the semantics (cf. Example 3), and returns the first event in that order.

Finally, the *Closure* computation, crucial for this procedure, is responsible for detecting event patterns. Given a pre-processed program where all event-rules are binary, *Closure* matches events to bodies of event rules, produces new temporary rules containing information about what events still need to occur to trigger an event pattern, and returns a new set of (complex) events that are true. During this procedure, the event component of the program is partitioned between *permanent rules* and *temporary rules*. Permanent rules have the form $o(e) \leftarrow body$ and come from pre-processing the original program. They can never expire and are always available for activation. Temporary rules have the form $o(e)\mathtt{OP} \overset{id_2}{\underset{id_1}{\Leftarrow}} body$ and arise from partially satisfying a permanent rule. They are valid only for some particular iterations (unless as we shall see, if their *ids* are not open). Then, they are either deleted (in case they are expired without being satisfied) or transformed (if they are partially satisfied). We say these rules are expired if the difference between the current global *id* and the rule's ending *id* is greater than 1. Moreover, temporary rules also have the information about the operation OP which defines the constraints needed to satisfy an event pattern.

Definition 14 (Closure).

Input: $ESet, P$
Output: $ESet', P'$
repeat {
Define $ESet' := ESet$, $P' := P$

Base Cases:

1. If $h \leftarrow o(e) \in P'$ **then:**
$ESet := ESet \cup \{h\{id_i, id_f\}\}$

2. If $h \otimes \overset{id_1}{\underset{id_2}{\Leftarrow}} o(e) \in P'$ **and** $id_i = id_2$ **then:**
$ESet := ESet \cup \{h\{id_1, id_f\}\}$ *and*
$P' := P' \setminus \{h \otimes \overset{id_1}{\underset{id_2}{\Leftarrow}} o(e)\}$

3. If $h \wedge \overset{id_1}{\underset{id_2}{\Leftarrow}} o(e) \in P'$, $id_1 = id_i, id_2 = id_f$ **then:**
$ESet := ESet \cup \{h\{id_i, id_f\}\}$ *and*
$P' := P' \setminus \{h \wedge \overset{id_1}{\underset{id_2}{\Leftarrow}} o(e)\}$

4. If $h \otimes \overset{*}{\Leftarrow} o(e) \in P'$ **then:**
$ESet := ESet \cup \{h\{^*id_i, id_f\}\}$ *and*
$P' := P' \setminus \{h \otimes \overset{*}{\Leftarrow} o(e)\}$

Negation Case:

1. If $h \leftarrow not(o(e_3))[o(e), o(e_2)] \in P'$ **then:**
$P' := P' \cup \{h \otimes \overset{id_i}{\underset{id_f^*}{\Leftarrow}} o(e_2), (h \otimes \overset{id_i}{\underset{id_f^*}{\Leftarrow}} o(e_2)) \neg \overset{id_i}{\underset{id_f^*}{\Leftarrow}} o(e_3)\}$

Operations Cases:

1. If $h \leftarrow o(e) \otimes o(e_1) \in P'$ **then:**
$P' := P' \cup \{h \otimes \overset{id_i}{\underset{id_f}{\Leftarrow}} o(e_1)\}$

2. If $h \leftarrow o(e) \wedge o(e_1) \in P'$ **then:**
$P' := P' \cup \{h \wedge \overset{id_i}{\underset{id_f}{\Leftarrow}} o(e_1)\}$

3. If $h \leftarrow o(e)\ ;\ o(e_1) \in P'$ **then:**
$P' := P' \cup \{h \otimes \overset{id_i}{\underset{id_f^*}{\Leftarrow}} o(e_1)\}$

Path Cases:

1. If $h \leftarrow path \otimes o(e_1) \in P'$ **then:**
$P' := P' \cup \{h \otimes \overset{*}{\Leftarrow} o(e_1)\}$

2. If $h \otimes \overset{id_1}{\underset{id_2}{\Leftarrow}} path \in P'$ **then:**
$ESet := ESet \cup \{h\{id_1, id_2^*\}\}$ *and*
$P' := P' \setminus \{h \otimes \overset{id_1}{\underset{id_2}{\Leftarrow}} path\}$

For each $o(e)\{id_i, id_f\} \in ESet$:
}*until $ESet = ESet'$*

For each: $rule_j \neg \overset{id_2 *}{\underset{id_1}{\Leftarrow}} o(e) \in P'$ and $o(e)\{id_i, id_f\} \in ESet$ **do:**

$P' := P' \setminus \{rule_j \neg \overset{id_2 *}{\underset{id_1}{\Leftarrow}} o(e), rule_j\}$

Return $ESet', P'$

Note that, the expression path is used to expand the interval where an event pattern holds, and is instrumental to define complex event operators as shown in [14]. E.g., $\mathbf{o}(e) \leftarrow \mathbf{o}(a.ins) \otimes \text{path}$ makes $\mathbf{o}(e)$ true in all paths $D_1 \xrightarrow{a.ins} D_2$ where $\mathbf{o}(a.ins)$ holds, but also in all paths obtained by expanding $D_1 \xrightarrow{a.ins} D_2$ to the right. Conversely, $\mathbf{o}(e) \leftarrow \text{path} \otimes \mathbf{o}(a.ins)$ makes $\mathbf{o}(e)$ true in all paths obtained by expanding $D_1 \xrightarrow{a.ins} D_2$ to the left. To cope with this, the procedure deals with the *open ids*: $*id$, $id*$ and $*$, where $*$ states that the right or left interval of an event pattern occurrence is unknown and $*id$ (resp. $id*$) states that the starting (resp. ending) of an event is any point before (resp. after) or equal to id. Since these ids can propagate to several events, we also say that $\forall id.\ id = *$, $id = *id_1$ if $id \leq *id_1$ and finally, $id = id_1^*$ if $id \geq id_1$.

Also, recall that the negation $\mathbf{not}(\mathbf{o}(e_3))[\mathbf{o}(e_1), \mathbf{o}(e_2)]$ can appear in the body of an event pattern rule as syntactic sugar for $\mathbf{o}(e_1) \otimes \neg(\text{path} \otimes \mathbf{o}(e_3) \otimes \text{path}) \otimes \mathbf{o}(e_2)$. Such an event starts when $\mathbf{o}(e_1)\{id_i, id_f\}$ is added to $ESet$, and then we add two rules to the program: $h \otimes \overset{id_f^*}{\underset{id_i}{\leftarrow}} \mathbf{o}(e_2)$ and $(h \otimes \overset{id_f^*}{\underset{id_i}{\leftarrow}} \mathbf{o}(e_2))\neg \overset{id_f^*}{\underset{id_i}{\leftarrow}} \mathbf{o}(e_3)$. The former rule says that the not-event becomes true when $\mathbf{o}(e_2)$ appears in the $ESet$. The latter rule checks if $\mathbf{o}(e_3)$ appears in the $ESet$ (before $\mathbf{o}(e_2)$), and in that case, the first temporary rule is removed, so that a later occurrence of e_2 does not make the not-event true. Importantly, the removal of temporary rules that arise from such negation patterns is performed after the fixed point is achieved, separating the monotonic construction of the $ESet$, from the non-monotonic behavior of the rule $\neg\leftarrow$.

Theorem 1 (Soundness and Completeness of the Procedure). *Let P be a program, π a path, and ϕ a transaction formula. $P, \pi \models \phi$ iff $P, \pi \vdash \phi$*

5   Discussion and Final Remarks

Our work can be compared to solutions that deal with the detection of events, and with the execution of (trans)actions. Event Processing (EP) systems, e.g. [1,4,23], offer very expressive event algebras and corresponding procedures to efficiently detect complex event patterns over large streams of events. As shown in [14], $\mathcal{TR}^{ev}$ can express most event patterns of SNOOP [1] algebra, failing only to translate the expressions requiring the explicit specification of time. Our procedure is inspired on the one from ETALIS [4] algebra, namely the idea of rule binarization, and program transformation. ETALIS has some roots in $\mathcal{TR}$, sharing some of its syntax and connectives, and a similar translation result can be achieved for this algebra (but omitted for lack of space). However ETALIS, as all EP systems, does not deal with the execution of (trans)actions. As such, one can see the procedure presented herein as an extension of ETALIS algorithm with the ability to execute transactions in reaction to the events detected. Moreover, EP-SPARQL [3] is an interesting stream reasoning solution, based on ETALIS, that provides a means to integrate RDF data with event streams for the Semantic Web. We believe that with the correct oracles instantiations, a similar behavior could be achieved in $\mathcal{TR}^{ev}$, and leave this as a future direction.

Several solutions based on action theories exist to model very expressively the effects of transactions that react to events, as [5,7]. However, these are based on active databases, and events are restricted to simple actions like "on insert/delete", thereby failing to encode complex events as defined in EP algebras and $\mathcal{TR}^{ev}$. The work of [11] proposes a policy description language where policies are formulated as sets of ECA rules, and conflicts between policies are captured by logic programs. It ensures transaction-like actions if the user provides the correct specification for conflict rules. Yet, only a relaxed model of transactions can be achieved and it requires a complete low-level specification of the transaction conflicts by the user. In multi-agent systems, [12,15] propose logic programming languages that react and execute actions in response to complex events. Unfortunately, actions fail to follow any kind of transaction model.

Several ECA languages have been proposed in the literature like [2,10,11] with very expressive event and action algebras. However, ECA languages normally do not allow the action to be defined as a transaction, and when they do, they lack from a declarative semantics as [19]; or they are based on active databases and can only detect atomic events defined as insertions/deletes [16,24]. As shown in this paper, with the appropriate instantiations of the *choice* function, $\mathcal{TR}^{ev}$ can be used as an ECA language where the action is guaranteed to execute as a transaction, and offer different operational behaviors depending on the application needs. Moreover, the procedure presented herein gives an important contribution to implement such an Event-Condition-Transaction language, and closing the existing gap to use it in real scenarios.

References

1. Adaikkalavan, R., Chakravarthy, S.: Snoopib: interval-based event specification and detection for active databases. Data Knowl. Eng. **59**(1), 139–165 (2006)
2. Alferes, J.J., Banti, F., Brogi, A.: Evolving reactive logic program. Intelligenza Artificiale **5**(1), 77–81 (2011)
3. Anicic, D., Fodor, P., Rudolph, S., Stojanovic, N.: EP-SPARQL: a unified language for event processing and stream reasoning. In: WWW 2011, pp. 635–644 (2011)
4. Anicic, D., Rudolph, S., Fodor, P., Stojanovic, N.: Stream reasoning and complex event processing in etalis. Seman. Web **3**(4), 397–407 (2012)
5. Baral, C., Lobo, J., Trajcevski, G.: Formal characterizations of active databases: part II. In: Bry, F. (ed.) DOOD 1997. LNCS, vol. 1341, pp. 247–264. Springer, Heidelberg (1997)
6. Behrends, E., Fritzen, O., May, W., Schenk, F.: Embedding event algebras and process for ECA rules for the semantic web. Fundam. Inf. **82**(3), 237–263 (2008)
7. Bertossi, L.E., Pinto, J., Valdivia, R.: Specifying active databases in the situation calculus. In: SCCC, pp. 32–39. IEEE Computer Society (1998)
8. Bonner, A.J., Kifer, M.: Transaction logic programming. In: ICLP, pp. 257–279 (1993)
9. Bonner, A.J., Kifer, M.: Results on reasoning about updates in transaction logic. In: Kifer, M., Voronkov, A., Freitag, B., Decker, H. (eds.) Dagstuhl Seminar 1997, DYNAMICS 1997, and ILPS-WS 1997. LNCS, vol. 1472, p. 166. Springer, Heidelberg (1998)

10. Bry, F., Eckert, M., Patranjan, P.-L.: Reactivity on the web: paradigms and applications of the language xchange. J. Web Eng. **5**(1), 3–24 (2006)
11. Chomicki, J., Lobo, J., Naqvi, S.A.: Conflict resolution using logic programming. IEEE Trans. Knowl. Data Eng. **15**(1), 244–249 (2003)
12. Costantini, S., Gasperis, G.D.: Complex reactivity with preferences in rule-based agents. In: RuleML, pp. 167–181 (2012)
13. Fodor, P., Kifer, M.: Tabling for transaction logic. In: ACMPPDP, pp. 199–208 (2010)
14. Gomes, A.S., Alferes, J.J.: Transaction Logic with (complex) events. Theory and Practice of Logic Programming, On-line Supplement, **14** (2014)
15. Kowalski, R., Sadri, F.: A logic-based framework for reactive systems. In: Bikakis, A., Giurca, A. (eds.) RuleML 2012. LNCS, vol. 7438, pp. 1–15. Springer, Heidelberg (2012)
16. Lausen, G., Ludäscher, B., May, W.: On active deductive databases: the statelog approach. In: Kifer, M., Voronkov, A., Freitag, B., Decker, H. (eds.) Dagstuhl Seminar 1997, DYNAMICS 1997, and ILPS-WS 1997. LNCS, vol. 1472, p. 69. Springer, Heidelberg (1998)
17. Margara, A., Urbani, J., van Harmelen, F., Bal, H.E.: Streaming the web: Reasoning over dynamic data. J. Web Sem. **25**, 24–44 (2014)
18. Müller, R., Greiner, U., Rahm, E.: AgentWork: a workflow system supporting rule-based workflow adaptation. Data Knowl. Eng. **51**(2), 223–256 (2004)
19. Papamarkos, G., Poulovassilis, A., Wood, P.T.: Event-condition-action rules on RDF metadata in P2P environments. Comp. Netw. **50**(10), 1513–1532 (2006)
20. Ren, Y., Pan,J.Z.: Optimising ontology stream reasoning with truth maintenance system. In: ACM CIKM, pp. 831–836 (2011)
21. Rinne, M., Törmä, S., Nuutila, E.: SPARQL-based applications for RDF-encoded sensor data. In: SSN, pp. 81–96 (2012)
22. Sheth, A.P., Henson, C.A., Sahoo, S.S.: Semantic sensor web. IEEE Internet Comput. **12**(4), 78–83 (2008)
23. Wu, E., Diao, Y., Rizvi, S.: High-performance complex event processing over streams. In: SIGMOD Conference, pp. 407–418. ACM (2006)
24. Zaniolo, C.: Active database rules with transaction-conscious stable-model semantics. In: DOOD, pp. 55–72 (1995)

Author Index

Printed in the United States
By Bookmasters